Value Stream 4.0

This book describes how material and information flows of a value stream can be synchronized with the aim of processing orders as quickly and waste-free as possible. The design of the value stream is systematically oriented toward the requirements of the respective business model. The approach described goes beyond the classic value stream method, which primarily improves material flow, and places its informational focus primarily on the release of production and transport orders.

Value Stream 4.0: A Systems View of Information Flow for Digitalization focuses on the material and information flow within a product family, beginning with the receipt of customer orders through all process steps until delivery of the finished product. It illustrates how the systematic accumulation and provision of information at the right time, place, amount, and quality can enhance efficiency. Additionally, it explores how information can be leveraged to enhance customer value.

This compact, yet powerful, book is written for production managers, lean managers, operations managers, researchers, and students.

Value Stream 4.0

A Systems View of Information Flow for Digitalization

Joachim Metternich, Tobias Meudt,
and Lukas Hartmann

CRC Press
Taylor & Francis Group
Boca Raton London New York

CRC Press is an imprint of the
Taylor & Francis Group, an **informa** business

Designed cover image: Joachim Metternich, Tobias Meudt, and Lukas Hartmann

First edition published 2026
by CRC Press
2385 NW Executive Center Drive, Suite 320, Boca Raton FL 33431

and by CRC Press
4 Park Square, Milton Park, Abingdon, Oxon, OX14 4RN

CRC Press is an imprint of Taylor & Francis Group, LLC

Originally published by Carl Hanser Verlag GmbH & Co. 9/16/22, in German, 9783446472297

ISBN: 978-1-032-95865-1 (hbk)
ISBN: 978-1-032-95831-6 (pbk)
ISBN: 978-1-003-58690-6 (ebk)

DOI: 10.1201/9781003586906

Typeset in Times
by codeMantra

Contents

Preface

Joachim Metternich, Tobias Meudt, and Lukas Hartmann

Today, the value stream method is the standard procedure for systematically improving the value creation processes of manufacturing companies. The method focuses on the total material flow in a product family – starting with how the customer order is received, procurement and transportation of raw and primary materials, followed by the manufacture, as well as pre- and final assembly, and finishing with the delivery of the finalized product. The declared goal of the method is to keep material "flowing". Therefore, within the framework of the value stream analysis, material inventories and their causes are particularly looked upon. The value stream design that follows the analysis tries to combine processes either without inventories or with the lowest possible material buffers. The credo here is: Reduced inventories = Shorter throughput times. Thus, by eliminating the causes of inventories, throughput times decrease. At the same time, the ability of a value stream to fulfill a new customer order increases. There are many publications and reference books on the value stream method, each with a specific technical focus (e.g., describing the interrelationships in the value stream through formulas) or addressing a specific type of production (e.g., small batch production). So why write another book about the topic? Why does this book exist?

As for now, the basis of all present contributions to the value stream method is the groundbreaking work by Mike Rother and John Shook, *Learning to See*. Based on this classic work, all subsequent works on the value stream method primarily address the material flows and their associated control information. However, after running many value stream mapping projects, we – the authors of this book – have found that insufficient attention is paid to the analysis of information flows. Missing material and unavailable machinery are usually not the only factors bringing an order to a halt. Meaning it either cannot be started, further processed, or completed. Missing information is often the cause: For example, to continue work, more clear product specifications could be needed, inquiries about a drawing are required, and machining or measuring programs went missing. Just like Mike Rother and John Shook, we also deem this information already included in the material flow. And if this information flow is not coordinated with the material flow, a value stream cannot achieve maximum performance. Therefore, analogous to material logistics, information logistics must also be systematically defined, ensuring that the right information is available at the right time, in the right quality, and at the right workplace. Regarding individual customer orders, fast response and short throughput times cannot be achieved solely by processing the associated material quickly. Significant proportions of time are already lost well before production, namely in clarifying the customer's requirements, in customization engineering, and in work preparation. For this reason, we are expanding the focus of the classic value stream method to include all those involved in order processing, including the customer.

When considering information flows, it becomes apparent that – analogous to the handling of material – waste can also occur in the handling of information. So far, this could not be systematically recorded and addressed within the framework of the value stream method. Therefore, we introduce an information-related concept of waste, which sharpens the view for information logistics types of waste.

However, the actual reason for this book was the steadily advancing digitalization of processes, which has gained further momentum under the keyword "Industry 4.0". With the increasing availability of information along value streams, the question of waste avoidance is no longer the only one. A contemporary value stream design must also be able to answer the following questions:

- How can the accumulated information be systematically used to improve processes, e.g., by identifying quality deviations or critical machine conditions?
- How can information be used to increase customer value?
- What have the material and information flows to accomplish to be able to successfully implement a business model?

For this reason, we call this method "Value Stream 4.0". We hope you enjoy reading it and wish you every success in implementing it.

Yours,
Joachim Metternich, Tobias Meudt, and Lukas Hartmann

About the Authors

Dr. Joachim Metternich heads the Institute for Production Management, Technology, and Machine Tools (PTW) at the Technical University of Darmstadt. His work at PTW primarily revolves around advancing lean production with digital technologies and developing learning factories for this purpose. After completing his studies in industrial engineering and earning a doctorate, he gained extensive experience at both domestic and international production sites of leading German companies where he was involved in numerous value stream projects. Prior to joining PTW, he held a position overseeing the global production system of a mechanical engineering firm.

Dr. Tobias Meudt is the manager of Corporate Strategy & Portfolio Management at JUWI. He worked as a postdoctoral researcher at the PTW Institute at TU Darmstadt. He has been working intensively on topics of lean production, Industry 4.0, and the digitalization of companies in practice for more than ten years. His experience from consulting projects and industry working groups has been incorporated into the design of the methods. He did his doctorate on Value Stream Analysis 4.0 and accompanied the further development of the Value Stream Method 4.0.

Dr. Lukas Hartmann is a team leader for Manufacturing Excellence at EDAG Production Solutions. In his former position, he was the head of the Center for Industrial Productivity (CiP) at the PTW Institute at TU Darmstadt. In his research work, he focuses on the digital transformation of manufacturing companies and the efficient design of value streams in Industry 4.0. Within his projects at EDAG Production Solution, he designs complete smart factories from the first vision until their realization.

1 A New Perspective on the Value Stream Method

1.1 OVERSIGHT OF THE ENTIRE ORDER PROCESSING

Have you ever found yourself in a situation where a newly created value stream map offers a detailed view of material flow but falls short in identifying the underlying reasons for uncontrolled inventory, quality problems, and long order processing times? If, in retrospect, you've wondered "Why?" and looked deeper into the functions of production, order planning, product development, eventually sales, product management, etc., to identify the root causes, then you've made a cognitive journey similar to that of the authors of this book. But let's start from the beginning.

Let's examine the lead times of customer orders using a hydraulic pump manufacturer as our example. In this case, only 4% of the total order time is dedicated to the production phase. The predominant portion of the lead time is allocated to engineering and parts procurement.

It's crucial to recognize that many companies spend a considerably larger amount of time on activities such as understanding customer requirements, clarifying design, and the preparation of production processes for new orders. These preparations encompass tasks such as creating detailed drawings, configuring machine programs, compiling essential material documentation, and formulating work plans among other responsibilities. Surprisingly, these activities consume a substantial amount of time, often more than the actual production phase itself. This invaluable time, which can significantly influence the success and timeliness of a customer order, is frequently expended in the so-called "indirect" areas outside the production scope (Figure 1.1).

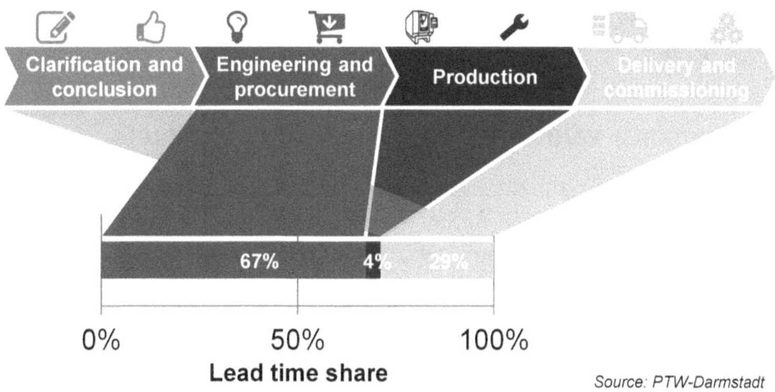

FIGURE 1.1 Lead time portions of the whole order processing.

DOI: 10.1201/9781003586906-1

Regrettably, the processes within the upstream indirect domains, which primarily handle information, often receive scant attention from the conventional value stream method. This oversight leads to untapped potential, particularly concerning the overall order lead time.

Conversely, techniques for analyzing indirect areas tend to be confined to these specific functions, disregarding downstream processes or the material flow within production. Examining processes within indirect areas is relatively challenging due to their non-physical nature. Here, information must be acquired, processed, and transmitted. Unlike direct areas, distinguishing value-adding activities from waste is more difficult and often goes unnoticed. Indications of inefficiencies in information flows typically emerge when information is either unavailable or not provided in the right quality. Numerous factors can impede the flow of information, and Value Stream Mapping 4.0 provides a means to comprehensively analyze and address these issues.

1.2 INFORMATION AS A CRITICAL PRODUCTION FACTOR

Many value stream maps are still created this way today: First, the process steps are drawn in, then the collected process data is entered under the process box, and then the stocks of material of preliminary and final products are recorded. Only at the end, the arrows for the information flows for controlling the value stream are considered. This is followed by the identification of improvement opportunities ("Kaizen flashes"). The value of information is reduced by this procedure to its good (or improvable) use for controlling material movements.

To make manufacturing companies and their value streams ready for the information age, we need to harness the full value of information in every possible dimension. Information properly integrated into the value stream can serve as the basis for a continuous improvement of processes or, given its connection to the physical product, enable additional services for internal and external customers. In addition to streamlining and digitalizing information flows, new potentials can thus be used to upgrade existing value streams. The value of information will increase in future, and thus information flows and their requirements in value streams should be looked at as a whole (Figure 1.2).

Below are some inspirational examples of value stream improvements which can be achieved using current concepts and technologies:

High-Performance Value Streams for Individual Products

Short delivery times in the manufacturing of individual products require lean information and material flows. Processing customer-specific information must be standardized and, if possible, automated. This can be achieved, for example, if a customer (her- or himself) carries out the individualization/parameterization of her or his product in an online configurator. This way, customization development can be transferred to the customer, and valuable processing time for adaptation engineering is saved. The subsequent processes are to be connected in such a way that information is transferred continuously to the machines without avoidable interactions with people (e.g., in work preparation). This concept has been implemented, among other things, in the production of individual pumps or punching tools. Here, customers can

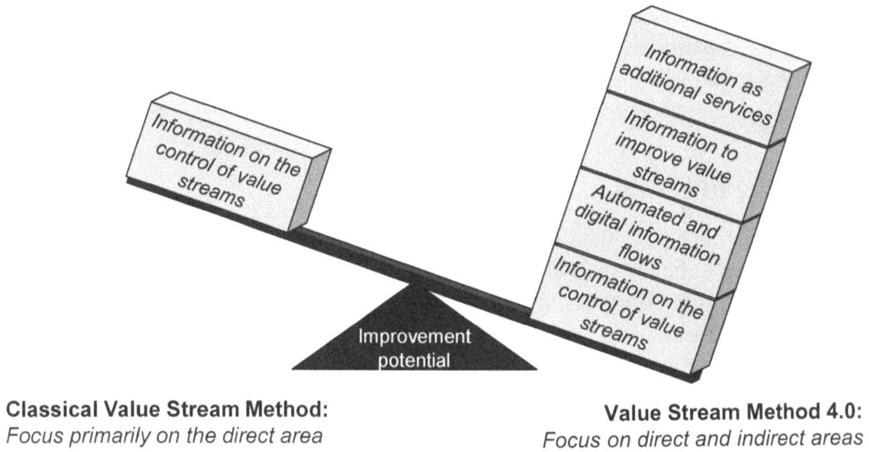

FIGURE 1.2 Focus on information within the classical VSM and the VSM 4.0.

configure a product in a predefined design space. The associated machine codes are generated automatically, and the orders with machine codes are sent digitally to the machine without an intermediate step via an engineering or design department. All necessary internal logistic processes are also digitally connected to avoid waiting times due to missing information and to enable the shortest possible lead time.

Identifying Deviations before they Become Problems

The goal is to collect information from processes, process it, and make decisions based on the information obtained. Artificial intelligence in the form of machine learning is used to evaluate data and detect deviations or failures before they lead to rejections, rework, or unplanned downtime. It is important to act instead of reacting (upgrading the poka-yoke principle).

This concept has already been implemented in many companies. One example is the retrofitting of a saw with sensors. With the recorded vibration and temperature data as well as failure information, an artificial neural network is then trained in software. The expected time of failure is fed back via a monitor, and a timely replacement of the saw blade is suggested to prevent damage.

Up-to-Date Information On-site via Assisting Systems

Worker assistance systems enable the provision of on-site information tailored to the needs of the staff and, if necessary, the ability to seek expert advice through video calls. One key advantage is the flexibility to update information right up until the activity is performed without requiring additional effort as in the case of paper-based work instructions.

For instance, consider a scenario where a company utilizes smartwatches to notify their internal logistic workers when the stock of electronic components for an SMD assembler is running low. This proactive approach allows the logistic workers to prevent any potential machine downtime caused by missing materials. Another example involves the use of data glasses during maintenance work on complex systems, which directly display action instructions to the personnel in their field of vision. This allows the workers to have unrestricted hand movement during tasks or seek assistance from experts when unexpected issues arise.

Technology Integration

Promising technologies, such as automated guided vehicles (AGVs) or cobots, have the potential to enhance the value stream by strategically optimizing specific tasks. The benefits can be observed through the reduction of simpler standard work or the decrease in lead time achieved by AGVs, which enable direct transportation routes from the source to the destination.

For instance, in automotive engineering, AGVs automatically transport tools for presses or in control cabinet manufacturing; they transport punching machines for setup processes. This automation aligns with the Single-Minute Exchange of Die (SMED) principle, where manual activities and associated control information are automated in these two cases.

Sustainability

By focusing on resource avoidance (such as minimizing electrical/thermal energy consumption and optimizing the use of raw materials and supplies) and promoting recycling, cost savings can be achieved.

For example, a manufacturer of plastic seals achieves cost savings by implementing 3D printing technology, which helps save expensive materials and minimizes the need for extensive machining. Similarly, a pneumatic valve manufacturer has implemented an energy monitoring system during the milling process. By measuring energy consumption, they can proactively trigger tool changes. This approach is based on the finding that worn-out tools lead to higher energy consumption and poorer component quality. As a result, the company has achieved reduced energy consumption, lower scrap rates, and a decrease in customer complaints.

1.3 OUR OBJECTIVE: ADVANCING THE VALUE STREAM METHOD

The value stream method, as described by Mike Rother and John Shook, is today seen as the "gold standard" in many companies aiming to enhance product flow, reduce inventories, and decrease lead times. It primarily focuses on the manufacturing value stream, which involves the smooth flow of parts and materials from raw materials to the end customer. Information flows are primarily considered from the perspective of production control and its improvement to achieve continuous production. This approach falls short when the objective is to fulfill individual customer orders quickly, flexibly, and efficiently. Merely considering production, material flows, and associated control information is insufficient for that case.

To address this limitation, the Value Stream Method 4.0 expands the focus of the classical value stream method to address all areas involved in order processing, including the customer. The value stream therefore encompasses the production

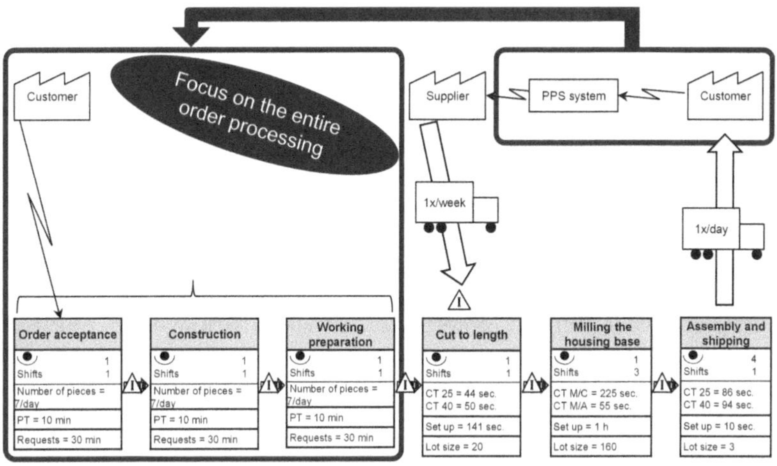

FIGURE 1.3 VSM 4.0 extends the focus to the entire order processing.

value stream and incorporates the processes of the indirect area within the order processing process. When depicting this expanded value stream in a value stream map, the previously summarized processes within the input arrows are "unfolded" and placed in front of the value stream. This provides a comprehensive overview of the entire value stream, from the customer order through the order processing in the indirect areas of the company and the production steps to the delivery of the finished product to the customer (Figure 1.3).

By adopting this broader perspective and aligning material and information flows, it becomes possible to eliminate waste, reduce throughput times, and enhance value creation efficiency.

Additionally, within the framework of this method, information is examined from three new perspectives:

- Waste in information handling;
- Utilization of information for process improvement;
- Utilization of information to enhance customer benefit.

The Value Stream Method 4.0 enables the creation of a lean value stream across the entire order processing process, starting from the initial customer contact in the administrative areas, through production, and ending in the shipment of the final product.

This method places its primary focus on analyzing and redesigning all essential information and material flows within a value stream. The information considered within the VSM 4.0 encompasses information required for production control, information to fulfill the tasks in every single process step and in the supporting functions, as well as utilization of information for various purposes, such as process improvement and services within the business model.

2 Setting the Stage for a Successful Project

2.1 FAMILY SELECTION AND VALUE STREAM IDENTIFICATION

For an effective value stream project, it is crucial to narrow down your focus to a defined product family, especially when dealing with a diverse range of products encompassing numerous variants and requirements. The relevance of a product family can be determined through various factors such as sales volume, profit, and manufacturing processes. From a value stream standpoint, a product family is a grouping of products that follow a sequence of processes sharing as much similarity as possible and utilizing the same resources. To delineate a product family and its corresponding value stream, the following three refined options are presented (Figure 2.1):

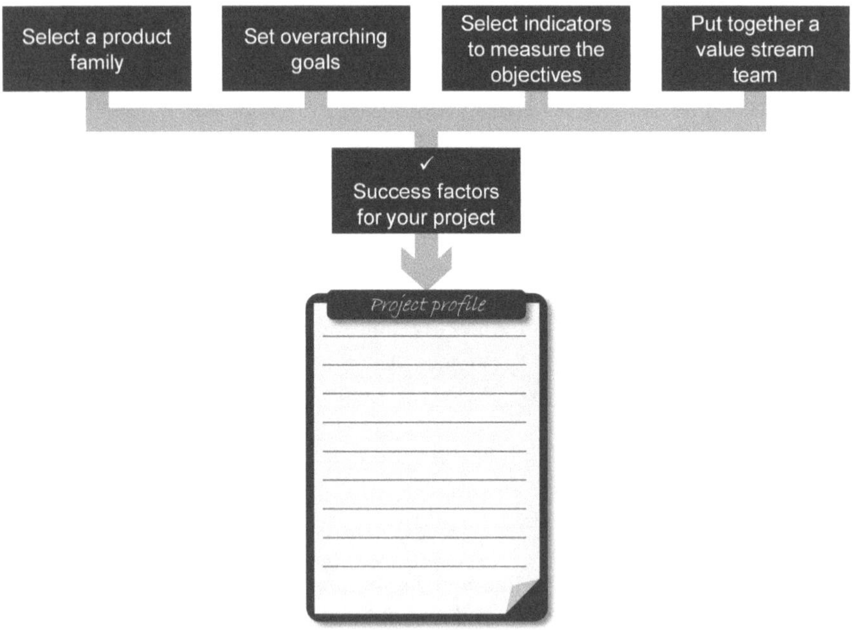

FIGURE 2.1 Project profile.

DOI: 10.1201/9781003586906-2

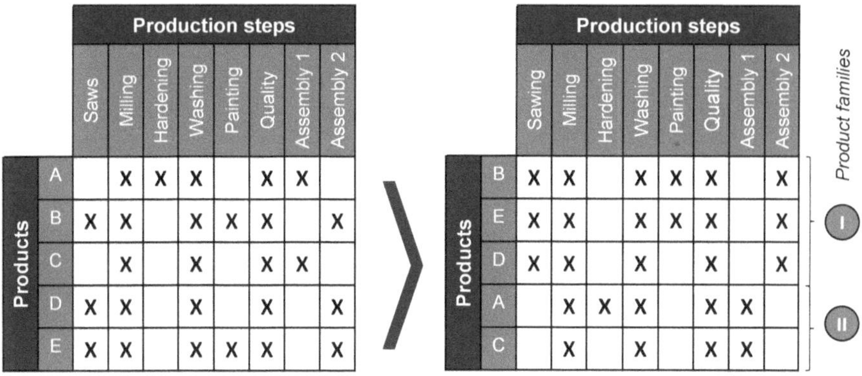

FIGURE 2.2 Product family matrix.

NOTE

When initiating your first project, choose a product family that holds significant importance to your company while ensuring the associated value stream is not overly complex. This approach enhances project visibility and reduces risks associated with assessing and managing a challenging project scope.

Option – Product Family Matrix[1]

This method is particularly effective when dealing with a limited number of products. Within a matrix, each product and necessary process steps are marked. Following the next step, product families are created by grouping products that share similar production steps. Determining which products belong to a family requires careful consideration to ensure the consolidation of similar production steps and resources. However, for large product families or extensive parts lists, this process can be time-consuming and get complex, especially when dealing with numerous article numbers (Figure 2.2).

Option – Production Flow

In this option, product families are established based on typical production sequences that individual products follow. A specific sequence is identified and chosen as a representative product family. Initially, products that follow the same processes as the selected sequence are not taken into consideration (Figure 2.3).

Option – Family Compilation

Another possibility is to categorize individual products according to certain features or combination of features. The following figure provides Examples of features categorized by raw materials, geometry, complexity, functionality, and handling (Figure 2.4).

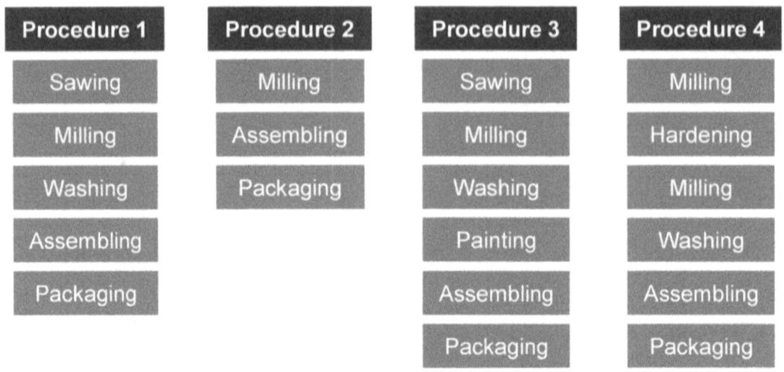

FIGURE 2.3 Production sequence.

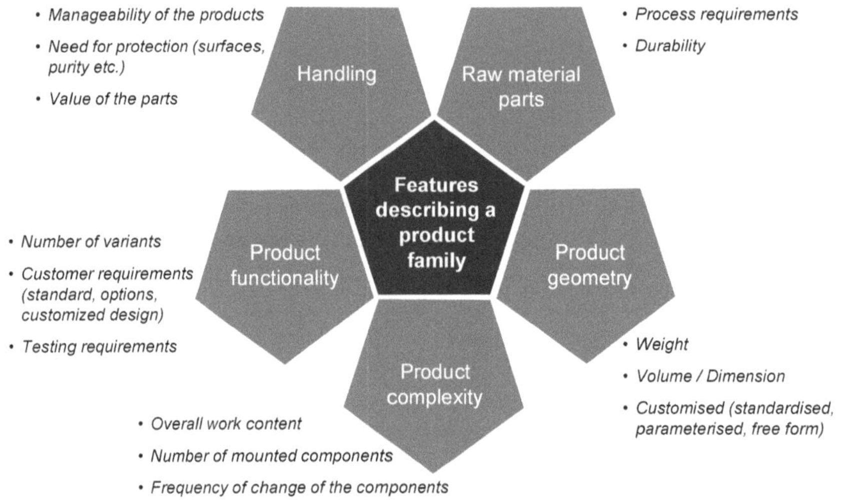

FIGURE 2.4 Family compilation.

2.2 SETTING OVERALL PROJECT GOALS

During the goal-setting phase of the value stream project, it is essential to establish a broader focus that goes beyond waste reduction and inventory management. Critical topics, overall targets, and focus areas of improvement should be defined for the value stream project together with the management. Additionally, defining overarching goals for the new value stream is crucial. The goals can be categorized as follows (Figure 2.5):

NOTE

In most companies, the answer to all the aforementioned questions is typically "YES". However, it is crucial to carefully balance the project's scope, level of knowledge, and available resources in order to attain the desired outcome. It is essential to identify and pursue specific project goals tailored to the organization's needs.

FIGURE 2.5 Goals for improvement (questions).

TABLE 2.1
Project Targets (Table with Example Values. This Table Can Be Updated Continuously during the Project)

Project Goals	Unit	Initial Value VSA4.0	Project Target VSD4.0	Current Values Implementation
Lead time	Days	23.5	7	14
Productivity	Piece/ (persons & week)	24	36	29.3

2.2.1 Procedure

As a result, the overall project goals of the management are quantified with key figures. It is advisable to list the objectives in a table and document them in the project profile (Table 2.1).

2.3 ESTABLISHING MEASURABLE PROJECT GOALS WITH KEY INDICATORS OF PROCESSES

Before starting the analysis, the value stream team is to select helpful KPIs to quantify the overall goals. The selected KPIs are then transferred to the project profile with their respective units. A selection of typical KPIs is listed here (Figure 2.6):

NOTE

The (manual) recording of many different key indicators gives the value stream team the feeling that they have "completely analyzed" the value stream. This involves a lot of effort and is at the expense of clarity, as the amount of information makes it difficult to interpret the value stream map. It is therefore helpful to limit oneself to a few meaningful indicators.

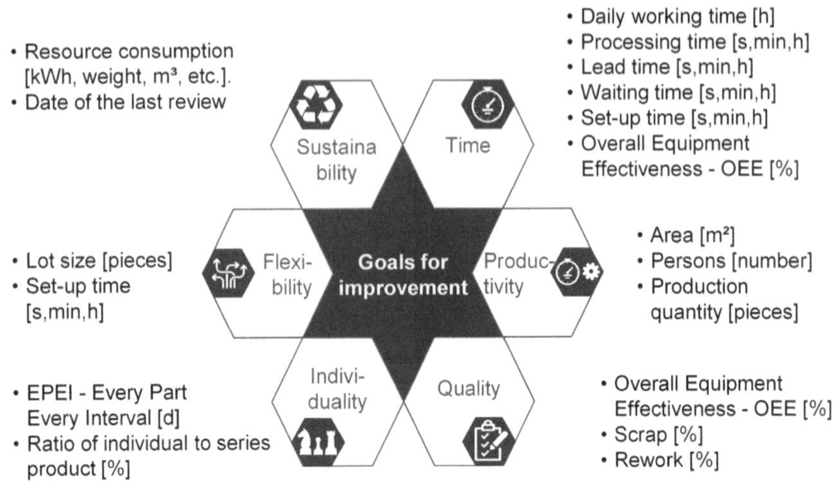

FIGURE 2.6 Goals for improvement (KPIs).

TABLE 2.2
Examples of KPIs for the VSA 4.0

Indirect Area	Direct Area	Logistics
• Processing time	• Cycle time	• Inventory
• Requests	• Failure rate	
• Number of pieces	• Set-up time	
	• OEE	
	• Requests	

2.3.1 Procedure

The desired KPIs for VSA 4.0 are recorded in the project profile, e.g.:

(If the collection of specific KPIs for a process does not add value, these KPIs are not calculated in the use of the method, e.g. the KPI "set-up time" if no setup exists in the process) (Table 2.2).

2.4 BUILDING A COMPETENT VALUE STREAM TEAM

The success of the value stream project not only depends on the knowledge and skills of the individual team members but also on the team roles. For your value stream project, it is useful to bring people from different departments and parts of the value stream into the team. This ensures that knowledge about specific elements from the sub-areas is available. Furthermore, people from departments like marketing, sales, and business development should also be included in the value stream project to incorporate requirements arising from new business models or the use of new technologies into the value stream design at the same time (Figure 2.7).

To ensure a successful project, certain roles need to be identified, trained, and assigned (Table 2.3).

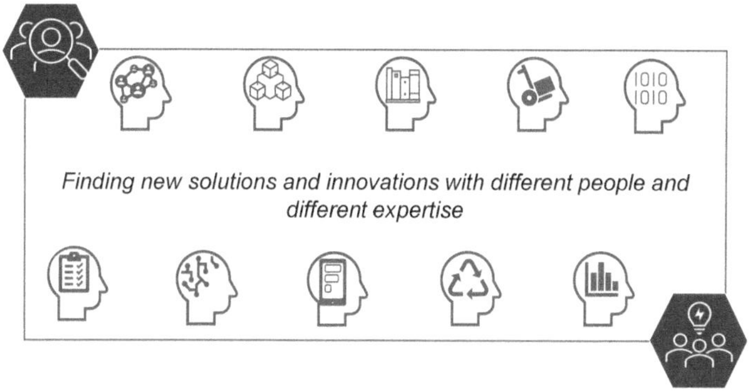

FIGURE 2.7 Value stream team.

TABLE 2.3
Roles within the Value Stream Team

Top Management	Methods Champion	Value Stream Manager	Value Stream Team & IT-Experts	Employees within the Value Stream
• Clearly highlights the importance of the project • Reviews progress • Actively removes barriers	• Methodological knowledge • Facilitates workshops • Coaching • Can be internal or external	• Personality for team leadership (reduces social distances and creates open exchange) • Project management and value stream responsibility (delegating tasks, target/actual comparison)	• Support the value stream manager • Conceptual knowledge • Multidisciplinary background • Critically question statements and results • IT experts for e.g. ERP requirements or machine connectivity	• Process experts • actively involve • Receive basic training on VSM4.0, Lean or digitalization • Use knowledge and skills in your value stream project for problem solving and improvement

2.5 SUCCESS FACTORS FOR YOUR VALUE STREAM PROJECT

Top Management Support[2]

Top management support is essential during implementation. It is important to explain that the "value stream as a whole" is to be understood, that project goals are related to the business strategy, that personnel are to be qualified within the project framework and that the project is not a threat. Communicate the project to stakeholders outside of the project team.

Put Together an Interdisciplinary Team Capable of Action

Choose motivated people from different fields of expertise in a value stream, form a team and empower them to make independent decisions. The team spirit is important.

Ensure Training of the Team

A sustainable value stream project not only improves the process of investigation but also the involved employees in their ability to recognize potential, solve problems, and improve processes – ideally during the project. The results of the project depend very much on the knowledge of lean management and digitalization.

Corporate Culture for Cooperation and Change

A supporting culture within your company not only makes communication with colleagues and determining the current state of the project easier, thus ensuring a more efficient implementation of new concepts. Likewise, the ideas of each member are heard, regardless of hierarchy. This supports group decisions and the realization of possibilities for improvement.

Adapt the Method to Your Project

If the theory is not applicable one by one to your value stream, an adapted solution can be developed with the creativity of your interdisciplinary team.

Pulling Together with Good Communication

Clear, permanent, and effective communication must be ensured for all people working in the value stream. Transparency creates trust.

Find Solutions Faster with IT Support

Experienced IT specialists can open new perspectives and contribute ideas for solving problems by asking questions. Sometimes small things, like setting up an access authorization might already improve the process.

Provide Resources

This includes human resources, budget (possibly for consultancy, training, or hardware/software), and time to redesign a value stream and implement a solution.

Change Must Be Managed

The right size of the value stream must be chosen (product family), key figures must be collected, and the Plan-Do-Check-Act-Process must be carried out consistently. This requires the presence on site.

Use of IT Tools and Software

Furthermore, the value stream team can use helpful software for recording and designing the value stream. This includes software for visualizing value streams and software for data collection or analysis.

2.6 DETERMINE PROJECT PROCEDURE AND PROJECT AGREEMENTS

The most important key points can be recorded in a written project profile. This documentation can always be referred to if problems or questions should occur. Before the project starts, it is useful to run through the value stream once, having short discussions with the most important participants and noting important processes before the profile is drawn up.

Example for the organization of a value stream project:

1. Set expected project objectives (determined by management)
    - "Output quantity is adjusted to customer demand. The value stream must deliver 2,800 parts in two shifts operation."
    - "Establishing a KPI-based continuous improvement process (CIP) in the value stream."
2. Focus of investigation (determine product family – Section 2.1)
    - The value stream is considered from sales to the shipped product. Purchasing is omitted.
    - Identified value stream (processes): Sales, design, CNC programming, drilling, assembling, and shipping.
3. Key Indicators to be achieved (Section 2.2 and 2.3)
    - We are improving lead time (target 50%, currently 14 days), first time right (target 99%, currently 75%), and space requirements (target $100m^2$, currently $400\ m^2$).

4. Team and roles in the team (compilation of a value stream team – Section 2.4)
    - Project management (CEO): Emilia M.
    - Value stream manager: Greta H.
    - Value stream team: Stefan S., Markus R., Jörg B.
    - External consulting: Carla M., Caroline M., Maja H.
5. Timetable (Figure 2.8)

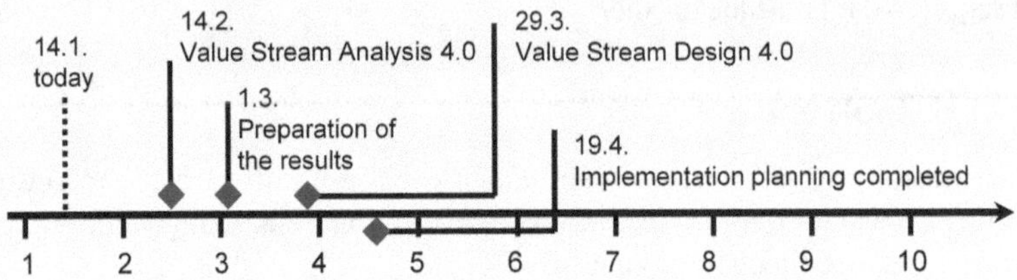

FIGURE 2.8 Example of a timetable for a value stream project.

6. Rules for cooperation
    - Decisions are made as a team, and every opinion is heard.
    - No parallel appointments.
    - Project time: Every Friday from 8 to 12 a.m.
7. Kick-off date and responsibilities
    - Best ideas seminar room, workshop material, and catering – Jörg B.
    - Project presentation – Emilia M.
    - Training – Carla M., Caroline M.

NOTES

1 Cf. here and all the options presented below: Erlach (2010).
2 Cf. here and in the following: Forno et al. (2014) and Shou et al. (2016).

REFERENCES

Erlach, Klaus (2010): *Wertstromdesign - Der Weg zur schlanken Fabrik*; 2., edit. und erw. ed.; Berlin, Heidelberg: Springer.

Forno, Ana Julia Dal; Pereira, Fernando Augusto; Forcellini, Fernando Antonio; Kipper, Liane M. (2014): Value Stream Mapping - a study about the problems and challenges found in the literature from the past 15 years about application of Lean tools; *International Journal of Advanced Manufacturing Technology* 72 (5–8), pp. 779–790.

Shou, Wenchi; Wang, Jun; Chong, Heap-Yih; Wang, Xiangyu (2016): Examining the critical success factors in the adoption of value stream mapping; *Proceedings 24th Annual Conference of the International Group for Lean Construction*, Boston, Massachusetts, pp. 93–102.

3 Value Stream Mapping 4.0 – Analyzing the Current State

3.1 PHASE I: CLASSIC WASTE ANALYSIS

The following figure 3.1 visualizes Phase I and II with all process steps.

In the following section, the concept of value creation and waste is elaborated (Section 3.1.1), followed by an explanation of the method steps of value stream mapping (Section 3.1.2) (Figure 3.2), and an example value stream is shown at the end (Section 3.1.3).

3.1.1 Value Creation and Waste

To conduct a VSA, it is important to understand and internalize the basic concept of value creation and waste. In general, anything a customer is willing to pay for can be considered as value-adding. Everything other than that is waste. All activities and production steps in the value stream can be divided into three categories:

- Value-adding activities (each and every activity for which customers are willing to pay),
- Necessary non-value-adding activities (e.g., loading and unloading a machine or gripping a tool) and
- Unnecessary non-value-adding activities/waste (e.g., long transport routes of tools needed but stored too far away from the place of value creation).

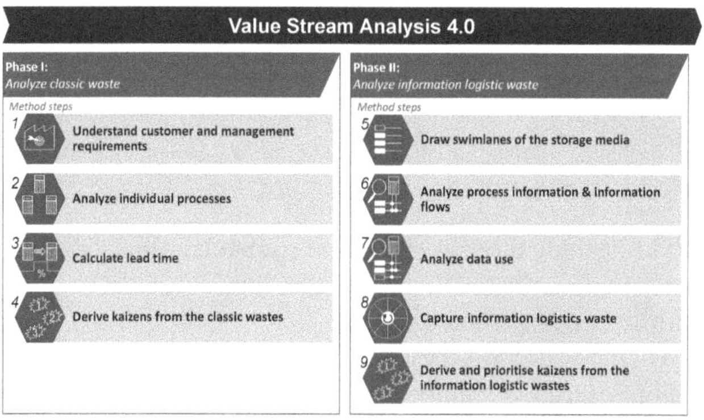

FIGURE 3.1 Value Stream Mapping 4.0 – method overview.

DOI: 10.1201/9781003586906-3

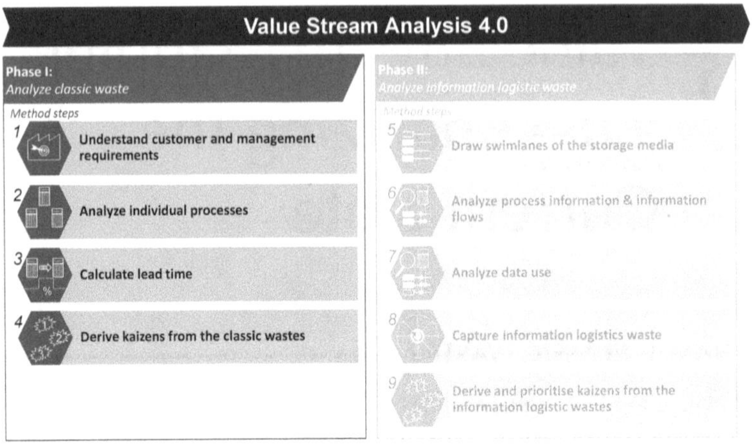

FIGURE 3.2 Value Stream Mapping 4.0 – Phase I.

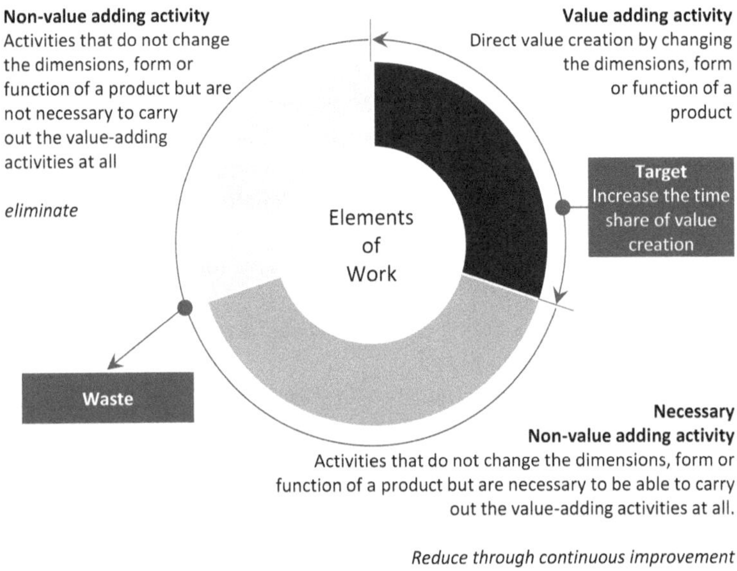

FIGURE 3.3 Elements of work.

Figure 3.3 shows just these three working time elements, which can be observed in a value stream mapping exercise. The overall goal here is to increase the value-added timeshare by eliminating or reducing waste.

Even though some waste is necessary to actually create value, these time components should be reduced to a minimum through continuous improvement. In general, waste and value-adding activities are not clearly separated from each other but

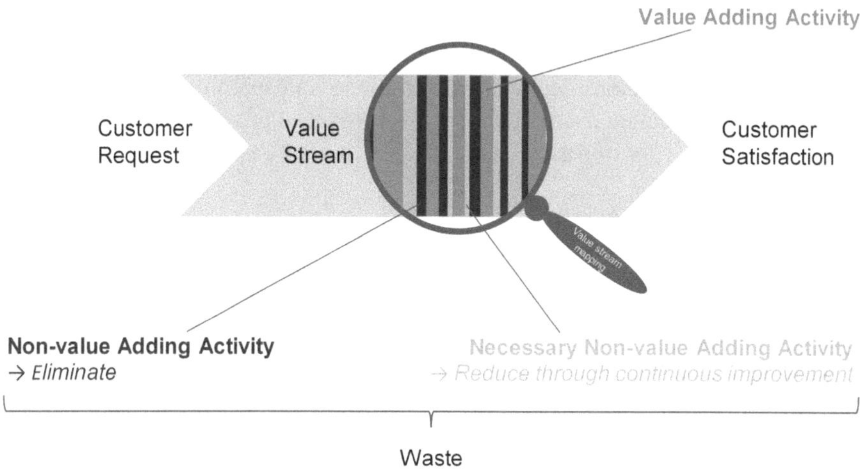

FIGURE 3.4 Value-adding -, non-value-adding- and necessary non-value-adding activities can often be found in a mixed sequence in a value stream.

can be found strung together in many different small-time shares within the value stream. The aim is to identify these individually (VSA) and eliminate them as good as possible (VSD). For process steps that are particularly significant for the value stream (e.g., bottlenecks), it is advisable to carry out a detailed workflow analysis while time-tracking and classifying each step into the categories of value-adding, necessary-non-value-adding, and waste. The workflow analysis is particularly not to be omitted where cycle-based work (e.g., in the context of an assembly line) is to be introduced or improved (Figure 3.4).

TAIICHI ŌNO BRIEFLY SUMMARIZES THE WAY OF THINKING ABOUT A VALUE STREAM

"All we are doing is to minimize the time between the moment we accept the order and the moment we collect the money. This we achieve by removing non-value-adding wastes". – Taiichi Ōno (Founder of the Toyota Production System)

The methods of lean production, which we will not discuss in detail here, support the reduction and avoidance of waste. A reduction of individual types of waste usually also leads to a reduction of other types of waste. Likewise, the occurrence of one type of waste usually leads to other types of waste.

3.1.1.1 The Classic Types of Waste

These classic 7 + 1 types of waste are briefly described below.

OVERPRODUCTION

Creating products more than customer demand and the premature completion of orders is called overproduction. Overproduced products can include or cause all other types of waste.

MOVEMENT

Unfavorable workplace design and provision of materials lead to unnecessary movement of people and materials.

REWORK AND DEFECTS

Improper handling or unstable processes can cause defects/rejects (cost). In the best-case scenario, rework can restore a product to the desired condition (effort and cost).

TRANSPORT

This is to be understood as unnecessary transport of products. This includes excessively long transport routes or the unnecessary storage of semi-finished products.

PROCESS OVER FULFILMENT

Producing products beyond a customer's requirement, e.g., a higher surface quality than required, represents an unnecessary expense.

WAITING TIME

Waiting time is usually related to the waiting of an employee. For cost-intensive systems, the time of a machine waiting for a person or missing material is considered.

INVENTORY

The unnecessarily high storage of finished goods, semi-finished products, purchased parts, and raw materials ties up capital and storage space, technology, management, and staff. In addition, there is the danger of products becoming obsolete.

UNUSED CREATIVE POTENTIAL OF STAFF

This occurs when well-trained employees only perform simple tasks and their skills are not used for continuous improvement.

3.1.2 Value Stream Analysis

Compared to the traditional value stream approach[1], the subsequent stages involve an in-depth analysis of processes, spanning from initial customer interaction (order), through both indirect and direct procedural value streams, all the way to the final shipping phase. The fundamental framework of the conventional value stream analysis (VSA), elaborated in the forthcoming sections, remains largely consistent. The modifications primarily pertain to the adjustments made in the visual illustrations of process boxes and logistics components.

Gain an initial comprehensive understanding by starting with a "quick run" through the whole value stream, spanning from the initial customer interaction to the final shipping phase. Subsequently, initiate the documentation process starting at the last process step and trace the flow upstream to the order acceptance stage, involving the initial customer contact. In this way, you can analyze the stream of each order and, most importantly, get in touch with everyone directly involved in the execution of the order.

Inquire the origin of order-related information and materials as well as their transfer. Focus on documenting the inventory levels of raw materials and partially finished products (and in indirect processes/office on the level of undone orders). Cases where materials accumulate or where there is significant stock and buffers in supermarkets indicate possible discrepancies in synchronization between the different stations within a value stream.

Employ "why" questions to dig into the underlying reasons for the existence of stockpiles. Uncovering these reasons will lead you to improvement opportunities, often called "kaizen" opportunities.

3.1.2.1 Step 1: Understanding Customer and Management Requirements

By precisely defining customer requirements for the value stream or product family, the value stream team gains sharper insights during on-site analysis compared to focusing solely on the customer takt-time (CT). Besides an analysis in a mass production environment, the initial step of calculating the customer takt-time – typical in classic value stream analysis – is occasionally omitted, especially when dealing with customized machinery or plant construction due to varying customer-specific workloads and demands. It is nevertheless recommended to try to calculate the customer takt-time even in these cases.

Outlined below are categories and questions essential for comprehensively describing customer and management requirements for the value stream to be analyzed. Depending on the company, additional categories might be needed to precisely capture the value stream's demands.

SERIES AND SMALL SERIES PRODUCTION – TAKT

The customer takt (CT) stands for the time allocated to the value stream for producing a unit of the specific product family. This value is obtained by dividing the working hours by the average demand within a given period.

Essentially, the customer cycle establishes the pace at which the value stream operates for a particular product.

$$\text{Customer Takt} = \frac{\text{Working Hours per Period}}{\text{Customer Demand per Period}}$$

Note: Working hours and customer demand must refer to the same period of time (e.g., per day, week, or month). Also, shared services/processes have to be considered.

Shift correction: If there are different number of shifts in the divisions, the CT is calculated individually for these divisions.

MACHINERY AND PLANT ENGINEERING – PROJECT BUSINESS AND TAKT

A distinguishing attribute of mechanical and plant engineering lies in the prolonged durations of processes within the indirect project sphere, particularly during order clarification with clients. This contrasts with the subsequent production and assembly phases in the direct areas, which tend to be quicker. There is often a lack of visibility into real-time progress or setbacks in projects, and working on numerous projects simultaneously presents a challenge for seamless prioritization and execution.

When characterizing a value stream, the following insightful queries can prove beneficial:

- Do the product variants fundamentally serve the same purpose?
- What factors determine the variants (size, material, color, etc.)? Do these factors only have to be considered once within a process, or do they also have an influence on downstream phases?
- To what extent are "customized assemblies" standardized or parameterized?
- What is the average demand (takt) over a specific timeframe?
- Are various order management principles employed (e.g., Critical Chain Project Management (CCPM), MRP, etc.)?

DEMAND CHANGE AND INSTABILITY

Companies rarely experience uninterrupted demand for their products. Instead, specific challenges arise that impose distinct requirements on value streams:

- Does demand exhibit growth or decline?
- Are there seasonal demand fluctuations or variations prompted by trade shows?
- What frequency of last-minute changes is initiated by customers?

SPECIAL CUSTOMER REQUIREMENTS

Filtering out and categorizing different customer needs can provide valuable insights. The following questions help to create clarity:

- What distinct customer groups exist?
- Do individual customer(-group) needs vary?
- Which product attributes significantly impact quality?
- Can (pre-)products be segmented into top sellers and unique products so that uninterrupted operation of certain value stream segments is possible in the context of individualized requirements?

URGENT ORDERS, REPAIRS, SHARED RESOURCES, ETC. IN MAIN VALUE STREAM

Value stream mapping captures a momentary depiction of the product family's order fulfillment process. Beneficial insights arise when categorizing this product family's order volume relative to the company's total workload:

- Do shared resources exist within the analyzed value stream?[2]
- Does the value stream handle repair orders? What is the repair-to-order processing ratio?
- To what extent is there planning stability? For instance, are weekly schedules set in place?
- What proportion of express orders exists in comparison to orders that can be planned?

FUTURE CHALLENGES

When potential changes in the value stream are foreseen, proactively discussing them within the team is beneficial. Key queries to consider include:

- What implications are there for quality (rejection and rework rates, customer complaints)?
- Might a quality initiative be warranted?
- Does the integration of novel technology into products necessitate additional process steps?
- Is there a need for cost reductions?

In the context of VSA4.0, the analysis involves analyzing information flows from customers into the value stream and vice versa.[3] The request addresses the way in which customer requirements and orders are documented (variants, dimensions, desired quantities, and delivery dates) as well as the data that is communicated to

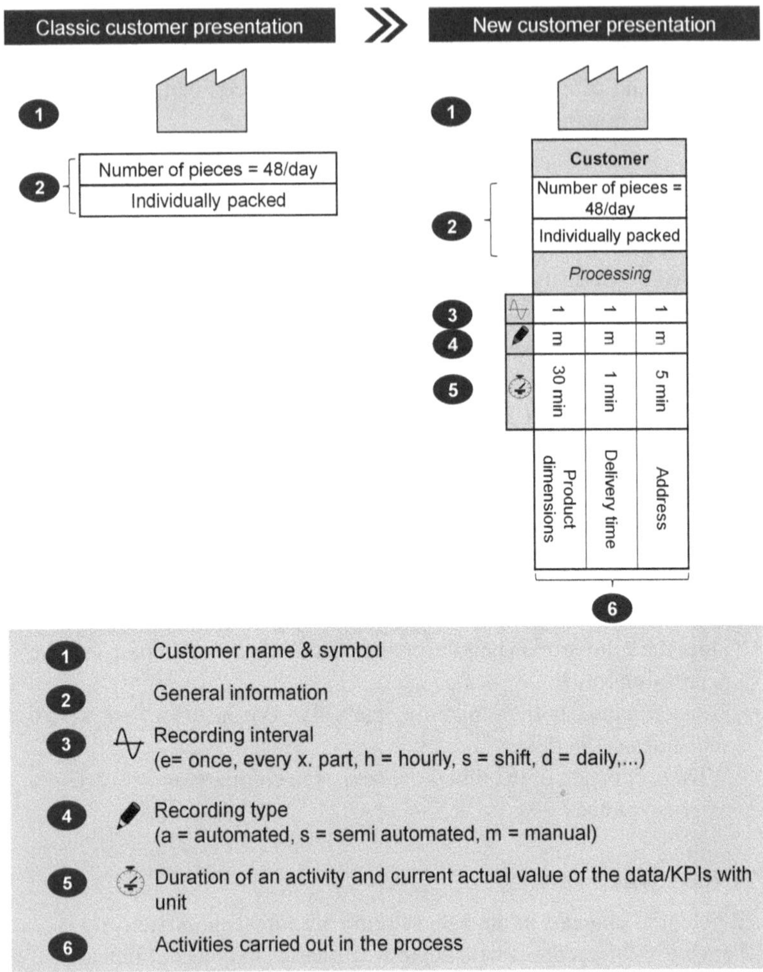

FIGURE 3.5 Introduction of new symbols – customers.

customers, such as product parameters or real-time order tracking. Fundamental data, familiar from classical VSA, is still captured within customer process boxes. VSA4.0 introduces a novel presentation for customer boxes, with individual customer information fields rotated by 90°, containing more comprehensive order processing details (Figure 3.5).

3.1.2.1.1 Procedure to Map the Current State

At the uppermost point of the value stream map, the customer is represented using a factory symbol. Supplementary details like delivery frequency or items per package are recorded in a data box below. Unlike the conventional VSA, *the value stream map now incorporates two instances of the customer*: preceding the order and

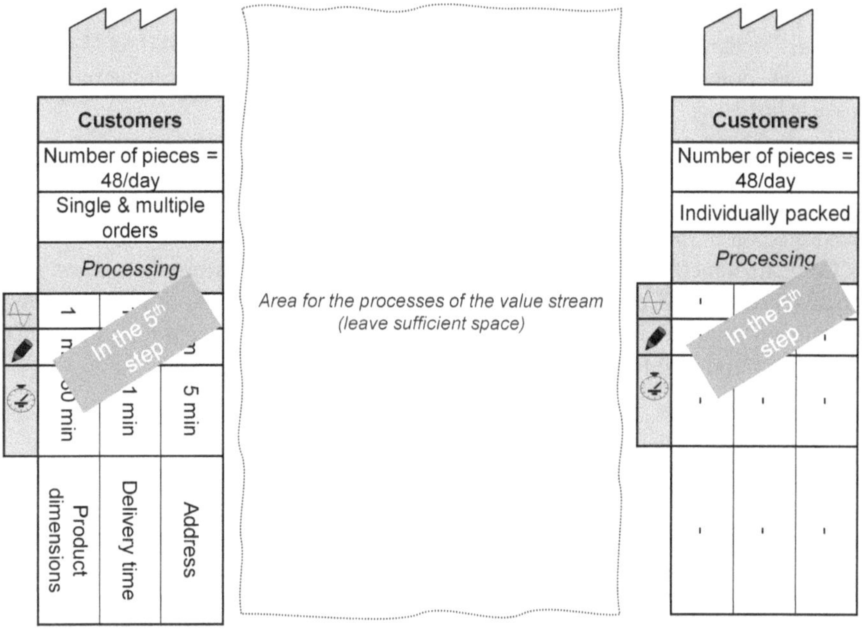

FIGURE 3.6 Value stream map up to the current step – suppliers and customers.

following the delivery. This dual representation allows for a comprehensive depiction of both the ordering process and the shipping and receiving stages at the customer's location (Figure 3.6).

To illustrate the method steps this book employs a practical example - Pumpen AG. To enhance clarity, we present selected segments from Pumpen AG's value stream that align with each specific method step. At the conclusion of both sections (Value Stream Mapping and Value Stream Mapping 4.0), the comprehensive value stream map of Pumpen AG is provided.

3.1.2.2 Step 2: Analyzing Individual Processes

The order fulfillment process includes several steps in indirect and direct areas, which can be divided or grouped into sub-processes. These sub-processes are visually represented using process boxes. Before the start of the value stream mapping, it is essential to determine the appropriate level of detail for representation in collaboration with the value stream team. An approach centered on tasks is recommended: Join process boxes with a distinct task-oriented focus, yielding well-defined outcomes. These outcomes could encompass an: order clarification's requirement list (internal sales), product design (development/engineering), or the tested, deliverable product (final assembly). These sub-process outcomes offer insights into whether projected project goals can be realized at this stage. For instance, it might be strategic initially to consolidate an entire department into a single process box, despite its inclusion of multiple sub-processes. In another project, this "department process box" can be detailed to grasp material and information flows, potentially presented as a distinct value stream with multiple individual process boxes.

A novel aspect of VSA4.0 is the revised depiction of process boxes (Figure 3.7) and logistics boxes (Figure 3.8). Data fields are rotated 90°, offering deeper insights into a process step or the logistics box itself. The specifics of this novel approach are elaborated on in the following text and visualizations.

In the fifth method step, we'll illustrate the connection of information flows from process boxes to information sources and sinks. However, our immediate focus remains on material flow analysis and the corresponding classical forms of waste. The exploration of information logistics waste will follow later.

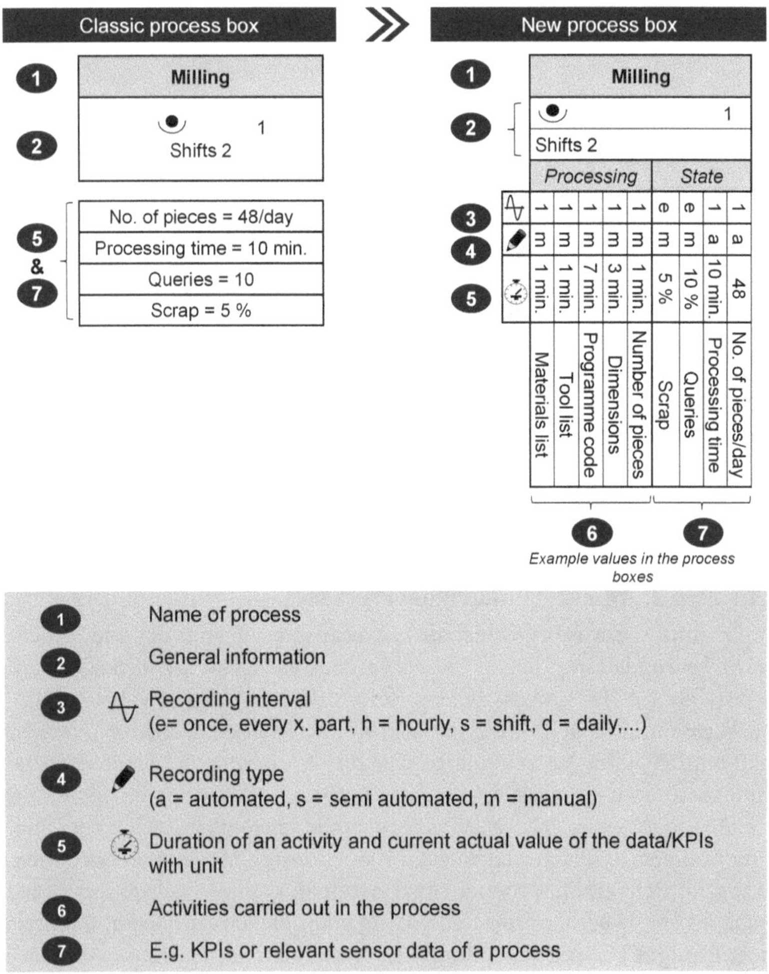

FIGURE 3.7 Introduction of new symbols – process box.

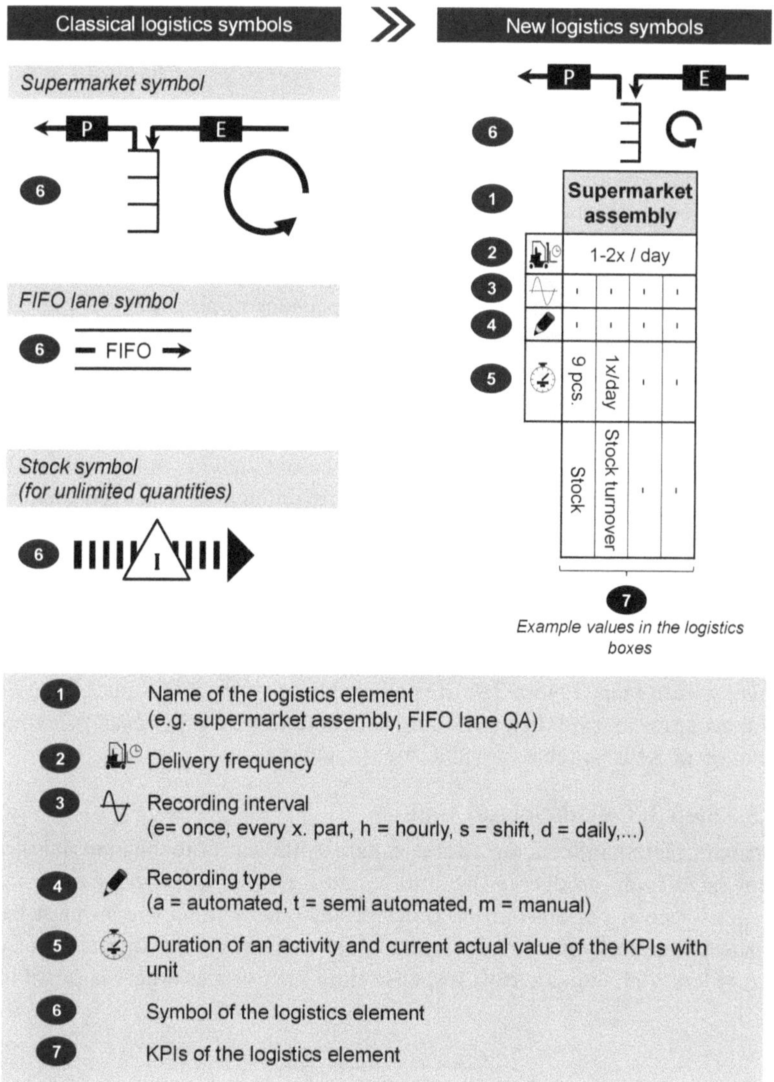

FIGURE 3.8 Introduction of new symbols – logistic symbols.

3.1.2.2.1 Procedure to Map the Current State

For every process, a process box is depicted using the VSA4.0 notation, linked to logistical elements, and control information flow arrows are inserted (refer to the illustration below).

External Material and Information Flows of suppliers and customers are depicted by arrows. The nature of the information flow arrows signifies whether ordering processes entail analogue or digital communication (e.g., customer product orders or supplier material orders). Supplementary details enhance process comprehension by capturing transport mode (truck- or air-freight), container dimensions, or frequency (daily or monthly). Such information is documented directly on the value stream map.

Internal Material and Information Flows illustrate the existing control mechanisms of order processing and the interconnections between individual processes concerning material and information logistics. Achieve this by, firstly, denoting centrally managed work orders with arrows extending from production planning and control (PPS) to the respective process boxes. Additionally, establish connections between processes and logistics elements (FIFO lanes, supermarkets, or stocks) on the value stream map, adding data about the frequency of planning.

Inventory is counted between two processes. To make this possible, the current stock of orders or materials must be gathered and noted in the logistics process box. Multiple storage locations are often involved between processes. It is crucial to determine where staff fetch or receive orders/products, aggregating all inventories consequently. In cases where multiple orders have been initiated but not completed within a process, this is regarded as inventory and highlighted as a Kaizen opportunity. Reasons include missing parts or unresolved information needing clarification.

Collecting Process Data On-site is crucial to document KPIs to provide a comprehensive process depiction. Independent counting or measurement of these values is essential, as it frequently reveals additional process inefficiencies during recording. The current KPI measurements are entered in the corresponding process box on the value stream map (Figure 3.9). It is advisable to verify KPI values from IT systems for accuracy and validity, extending to inventories as well. Chapter 2 contains a compilation of KPIs suitable for value stream mapping.

3.1.2.3 Step 3: Calculate Lead Time

The primary factors influencing a value stream's efficiency are the material throughput time, signifying production flexibility, and the order throughput time, indicating responsiveness. For customized order production, a value stream must balance both aspects: promptly meeting individual customer demands while efficiently processing orders. Achieving a short response time to customer orders is possible with

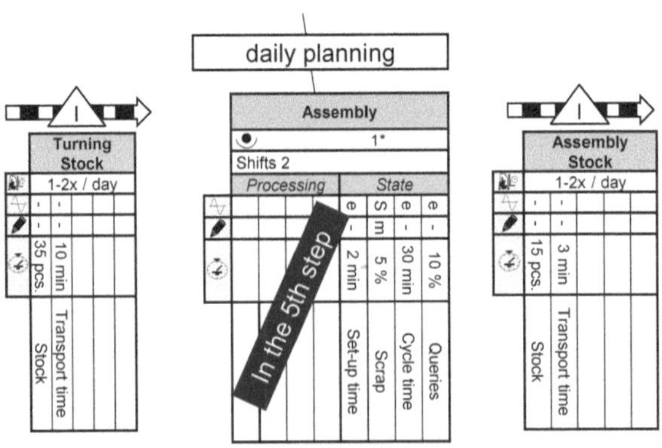

FIGURE 3.9 Value stream map up to the current step – process and logistic boxes.

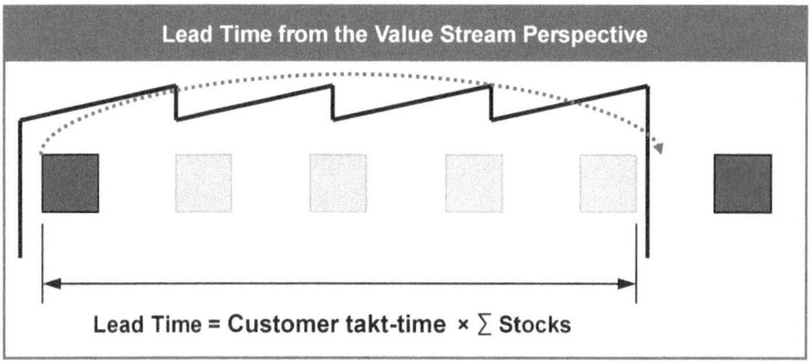

FIGURE 3.10 Lead time from a value stream perspective.

substantial finished goods stocks, but this comes at the expense of prolonged lead times and reduced flexibility concerning specific customer requirements.

3.1.2.3.1 Calculation of the Material`s Lead Time

Within the framework of the classic VSA, the material's throughput time stands as the utmost pivotal parameter. Its relatively straightforward calculation is founded on a simplifying premise: If a new material enters the value stream, it exits only toward the customer once all preceding components of that material group have traversed the value stream. On average, this point is attained when the entire stock of said material group completes the customer takt-time within the value stream. Hence, computing lead time involves multiplying the determined total material stock from the VSA by the relevant customer takt-time for that material (Figure 3.10).

$$\text{Material Lead Time} = \text{customer takt-time} * \sum \text{Stocks}$$

3.1.2.3.2 Calculation of the Value-Adding Timeshare

The value-adding timeshare signifies the portion of the overall processing duration that truly contributes value, enhancing the product from the customer's perspective. This involves dividing the total processing times by the overall lead time and presenting the result as a percentage. Generally, these values fall below 100%, which can serve as motivation to enhance the value stream design.

$$\text{Value-Adding Time Share} = \frac{\text{Total Processing Time}}{\text{Material Throughput Time}}$$

3.1.2.3.3 Calculation of the Order Lead Time

We now broaden our view to encompass the duration required by the value stream to complete orders for the specific product family in question. This order lead time explains the waiting period experienced by the customer, beginning from order placement. It is a key indicator of the responsiveness of the value stream. To achieve this insight, we must find out the average duration between order receipt and the dispatch of finalized orders. This necessitates the initial calculation of the order takt, signifying the time allocated to the value stream for executing a single order:

$$\text{Order Takt}_{\text{is}} = \frac{\text{Working Hours per Period}}{\text{Number of Completed Orders per Period}}$$

By utilizing the order takt and the inventory of pending orders for the specific product family in question, we proceed to calculate the average order lead time:

$$\text{Average Order Lead Time} = \text{Order Takt}_{\text{is}} * \text{Order Stock}$$

We can now evaluate the average time the value stream requires to complete a new customer order from the moment it is received. A comparison between the lead times of material and sales orders can offer interesting insights. If the lead time of the material significantly exceeds that of the customer orders, the reasons usually come from the semi-finished and finished goods warehouse. Conversely, if the material lead time constitutes only a fraction of the order lead time, then the substantial reaction time is expended outside production, such as in adaption development or production (Figure 3.11).

3.1.2.4 Step 4: Derive Kaizens from Classic Wastes

In concluding the initial phase of our approach, the value stream now encompasses the entire order processing, with the ability to document classic sources of waste. During this stage, the insights into the 7 + 1 waste types are to be incorporated into the value stream map.

3.1.2.4.1 Procedure to Map the Current State

Now, all the observed wastes ("Kaizen") need to be allocated to specific processes and consolidated. For instance, rather than naming individual wastes in the setup process, a Kaizen-like "prolonged setup times" is employed. This approach enhances the clarity of the value stream map (Figure 3.12).

3.1.3 Practical Example – Applying Value Stream Mapping

Pumpen AG specializes in crafting pumps for the chemical and food industries. Customer-specific needs like pumping capacity, chemical durability, hygiene, spatial constraints, and connectivity are gathered by the sales team. Standard parts are selected, or individual component designs are tailored to meet customer requirements. While Pumpen AG excels in delivering quality and reliability for personalized pumps, it has encountered a decline in the standard market segment recently. The Value Stream 4.0 initiative seeks to create a lean value stream for customized pumps, allowing Pumpen AG to markedly accelerate the delivery of custom-assembled "connecting flanges".

Currently, customer orders are received by the order acceptance (sales) department and logged into the IT systems. Subsequently, the engineering department customizes or even redesigns specific pump components according to customer

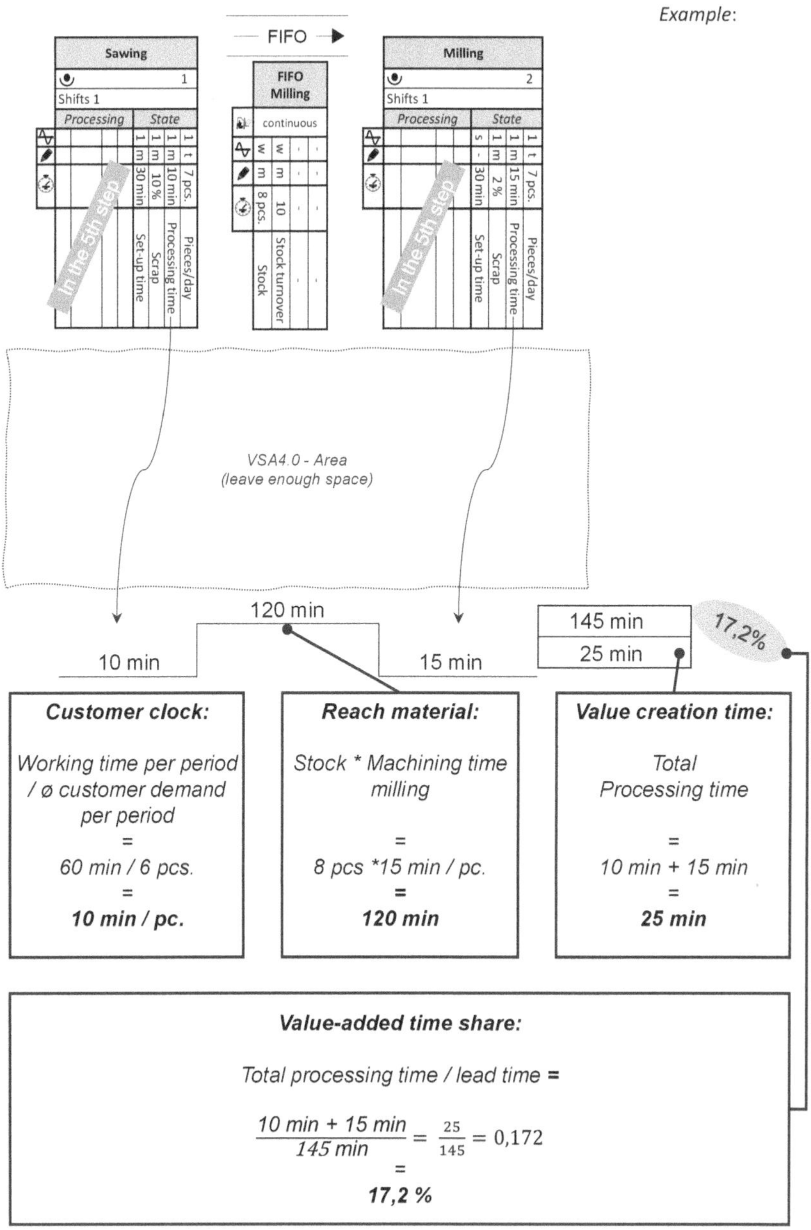

FIGURE 3.11 Value stream map up to the current step – calculation of classic value stream KPIs.

specifications. The following step involves adapting or reprogramming NC codes via CNC programming – the final subprocess in the indirect sector. Turning is the first production stage, crafting an individual flange from standardized components. In assembly, this flange is assembled into the pump housing. Conclusively, the pumps

Example:

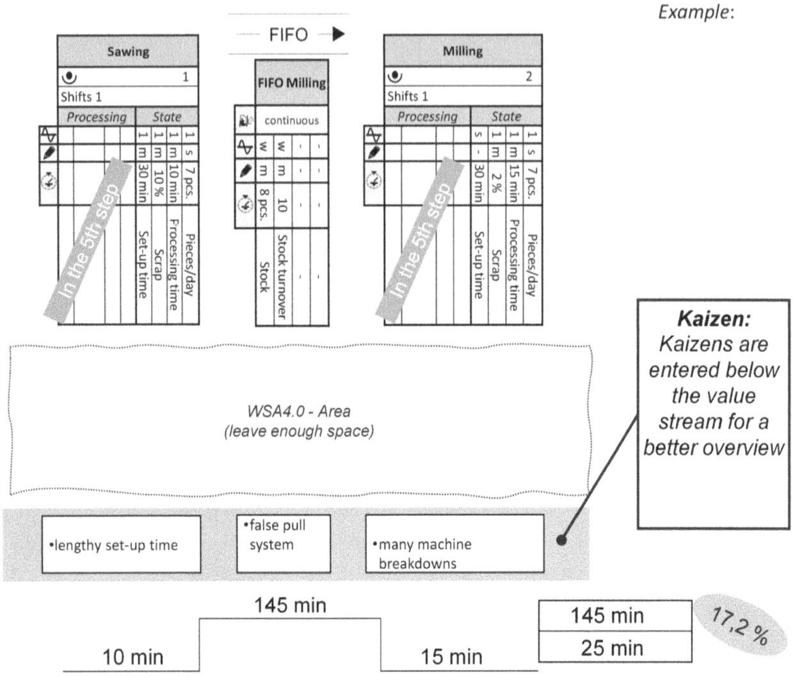

FIGURE 3.12 Value stream map up to the current step – information of Kaizen or Waste added.

undergo packing, labeling, and placement onto shipping shelves for logistics providers. This marks the end of the examined value stream. Thorough analysis of this value stream has been undertaken by the value stream team and captured within a value stream map (next page).

Identified opportunities for improvement (Kaizen):

- Noticeable differences are evident in the processing durations of individual processes.
- The value stream operates solely with unrestricted inventory (no pull system).
- The adherence to the chronological processing sequence based on delivery dates is inconsistent, with rush orders being overlooked on multiple occasions and order priorities frequently shifting.
- The value stream reveals instances of rejects, rework, and queries, resulting in recurrent disruptions (Figure 3.13).

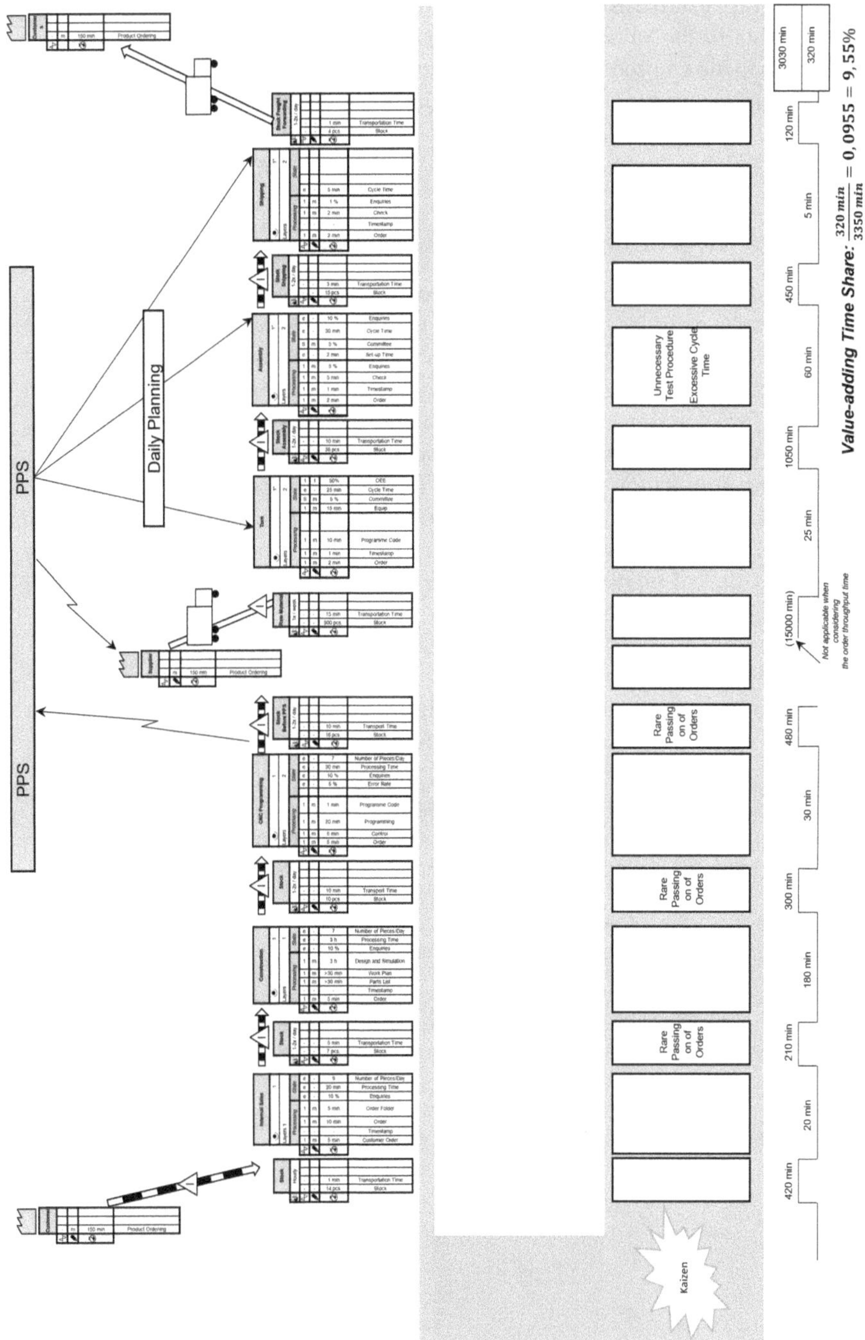

FIGURE 3.13 Value stream map example Pumpen AG.

3.2 PHASE II: INFORMATION LOGISTIC WASTE ANALYSIS

After concentrating on the material flow in the first phase, the information flow between the participants in the value stream is analyzed in the second phase. This begins with an introduction to the types of waste in information logistics (Section 3.2.1), followed by an explanation of the steps of the Value Stream Mapping 4.0 method (Section 3.2.2) (Figure 3.14). An example is used to illustrate the step-by-step application (Section 3.2.3).

3.2.1 Waste Types of Information Logistics

Information logistics includes the planning, control, implementation, and monitoring of all incoming (e.g., order information) and outgoing (e.g., measurement protocols) information flows within a specific process as well as the storage and arrangement of the associated data. Information logistics only includes those information flows that contribute to decision support.[4] Each information flow must therefore be assigned a main objective. The question should always be asked: "Why is this information flow existing?". The information may be required for specific tasks, for decision support, for process improvement or for the task itself. Once each information flow has been assigned an overarching goal, the evaluation of waste within the information flow can begin. A total of eight successive types of waste in information logistics are outlined below (Figure 3.15).[5]

3.2.1.1 Data Selection

A clear intent must be established for all chosen data/activities. The gathering or manipulation of data without a distinct purpose should be avoided. Beneficial inquiries include:

- Is the data's utilization clearly outlined (from a customer perspective)?
- Is a duration for data storage specified?

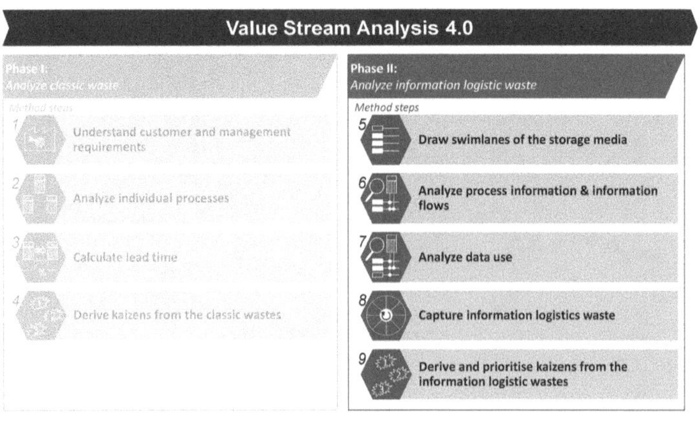

FIGURE 3.14 Value Stream Mapping 4.0 – Phase II.

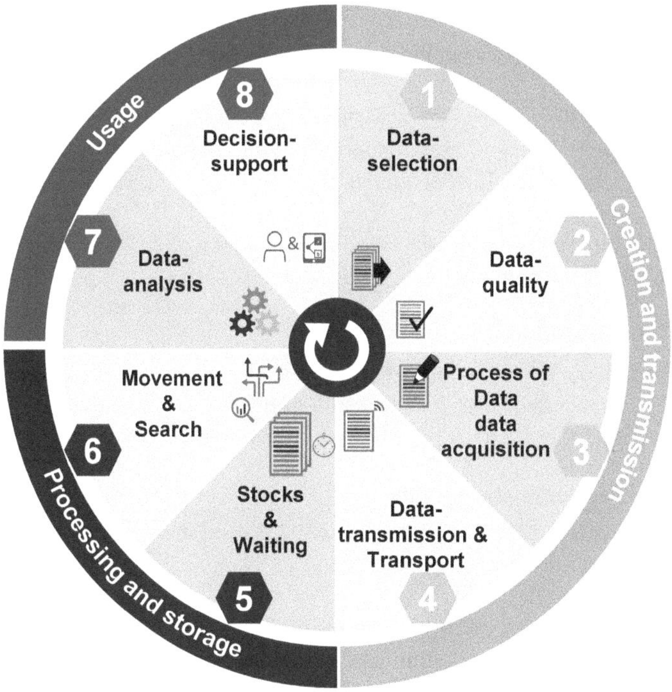

FIGURE 3.15 Information logistics waste types and concept of chronological identification.

3.2.1.1.1 Practical Examples

Relevant data remains unrecorded (e.g., KPIs for the morning discrepancy meeting), or there is data recorded that holds no relevance, etc.

3.2.1.2 Data Quality

Establish a measurable and replicable benchmark for data quality. Useful inquiries include:

- Do the frequency and level of detail of the data align with its intended use?
- Is it possible to correlate different sets of data?

3.2.1.2.1 Practical Examples

Mistaken entries or measurement inaccuracies, data captured at overly broad levels (e.g., pass/fail), incapability to link data from diverse sources, needless duplication of data collection, etc.

3.2.1.3 Data Collection Process

Develop a data collection process that strives for simplicity, standardization, automation, and accuracy. Beneficial queries include:

- Does the data collection strike a balance between costs and benefits?
- Is routine data collection automated?

3.2.1.3.1 Practical Examples

The data is noted down on paper and then manually transferred to computer systems.

3.2.1.4 Data Transmission and Transport

Minimize the employment of storage media and interruptions between them, prioritizing seamless continuity. Useful queries include:

- How, where, and through what media are data conveyed?
- Where is data stored, and does it have access control?

3.2.1.4.1 Practical Examples

Inaccessible data due to local data silos or paper-based storage (leading to a lack of transparency), inaccuracies in data transmission (e.g., due to interfaces), data loss during transmission, bottlenecks due to long transportation routes, and more.

3.2.1.5 Waiting Time and Stocks

The required data should be available before the procedure is initiated and processed promptly. Useful questions are:

- Is data accessible precisely when required?
- Are there waiting periods (human or machine-related) due to absent data?

3.2.1.5.1 Practical Examples

Delays in the measuring room or at test stands result in late transmission of test outcomes or measurement records. Excess data hinders efficient identification and prioritization of tasks, while unnecessary information occupies storage space.

3.2.1.6 Movement and Search

Searching for information is wasteful. Ideally, simple and automated provision of the required data at the right time is the key. Relevant inquiries include:

- Is data accessibility possible without the need for manual search?
- How can essential information be automatically linked with materials/orders?

3.2.1.6.1 Practical Examples

Workers leaving workstations for gathering documents. Delayed reporting of measurement outcomes requiring subsequent collection.

3.2.1.7 Data Analysis

Data should be precisely analyzed, harnessed, and tailored to specific objectives. Useful inquiries encompass:

- When was the data last used for analysis?
- Are suitable methods employed?
- Does analyzed data offer value?

3.2.1.7.1 Practical Examples

Findings from measurement areas and testing setups remain underutilized for comprehensive quality analyses across all components. Absence of basic statistical techniques like trend analysis.

3.2.1.8 Decision Support

Data should be tailored to users in suitable formats and mediums. Human roles in decision-making need clarification. Beneficial inquiries include:

- Is data prepared/visualized appropriately for its purpose?
- Can teams utilize information to enhance their processes?

3.2.1.8.1 Practical Examples

Inadequate chart types for issue visualization (e.g., missing trends or target directions). Information intended for management, not the decision-making team. Absence of processes (e.g., PDCA) for systematic information utilization.

3.2.2 Value Stream Mapping 4.0

The culmination of steps one to four leads to the value stream map, offering a comprehensive overview of the order flow and potential enhancements. These insights serve as a wellspring of ideas for subsequent value stream refinement.

Steps five through nine delve into each process step (i.e., process box) in the value stream. It assesses its upstream information sources, information processing, and downstream information usage. Concurrently, media utilized in the process are included and illustrated on the value stream map. This exact process provides a detailed representation of all essential communications within the value stream. Subsequently, information-logistic inefficiencies in individual processes are identified, leading to crucial improvement areas (information-logistic Kaizen). These Kaizens are addressed within the purview of Value Stream Design 4.0 (Chapter 5) or integrated into the ongoing daily continuous improvement process. The following pages explain the five phases of value stream mapping 4.0.[6]

3.2.2.1 Step 5: Draw Swimlanes of Storage Media

In this phase, document which storage media are used for the exchange of information within the value stream. Examples of storage media are IT systems (office applications, ERP/MES systems, etc.), paper-based materials (order slips, labels, drawings, etc.), and humans as information carriers. The objective here is to gain a thorough understanding of the actual storage media in use. The diversity of media encountered in specific instances can be quite surprising.

3.2.2.1.1 Procedure to Map the Current State

For each storage medium, draw horizontal lines, known as swimlanes, under the process boxes shown in the value stream map. Analog storage media (such as paper or human input) should be color-coded and categorized as "analog". Talk to employees in

Example:

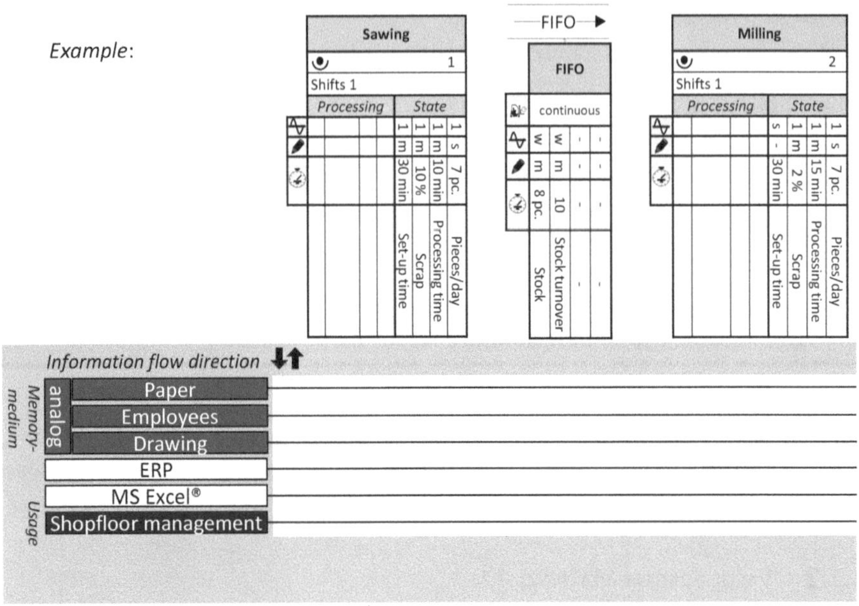

FIGURE 3.16 Value stream map up to the current step – storage media swim lanes added.

the field to find out what sources of information they rely on and where they store it. Maintain a comprehensive overview of the entire workspace and actively seek out information that is critical to the execution of tasks at each step of the process. This includes work instructions for various tasks and audit trails. Ask about the storage media used, whether it is paper-based or digitally available documents. Also, check whether defined key performance indicators (KPIs) are recorded in relation to the process status and establish the right link to a swimlane (if the KPI is tracked) (Figure 3.16).

3.2.2.1.2 Guiding Questions

- From where do you source information for your tasks (initiating an order, etc.)?
- What data is documented on which medium (storage medium) and how is it transferred?

3.2.2.2 Step 6: Analyze Process Information & Information Flows

This phase analyzes which storage media are used for the activities and KPIs of each specific process. For this purpose, vertical lines are drawn from an activity/KPI to the respective source or sink (swim lane). These connections are made using nodes that visualize media transitions (represented by multiple points on a vertical line).

3.2.2.2.1 Information Flow Procedure

When multiple sources/sinks are employed for data storage or retrieval, the corresponding number of nodes is depicted on the vertical line. If a process lacks a recorded KPI, no vertical line is drawn. In addition, arrows are inserted to indicate the direction of the information flow in order to clarify the transmission.

3.2.2.2.2 Detailed Analysis Procedure

Subsequently, a thorough analysis of activity and KPI data collection details is carried out. The actual time duration needed for each process step is documented. Furthermore, the frequency of KPI recording is recorded (e.g., weekly (w), daily (d), hourly (h), per piece (x), or once by the value stream analysis project (o)). Finally, for every activity/KPI, the data flow method is documented: manual (m), semi-automated(s), or automated data processing (a) (Figure 3.17).

3.2.2.2.3 Guiding Questions

- How do you execute your activities, and where does data originate from?
- Where is data stored, and how is data shared?

3.2.2.3 Step 7: Analyze Data Usage

Evaluate whether data collected from individual activities/KPIs is used for improvement processes or documentation. The aim of the procedure is to log and question the type of data usage, especially if this data is not directly used by another process. Other examples of data usage include process control, Shopfloor Management and the traceability of components. If further KPIs are derived from certain activities, these are marked off in a separate swimlane (calculation swimlane). For example,

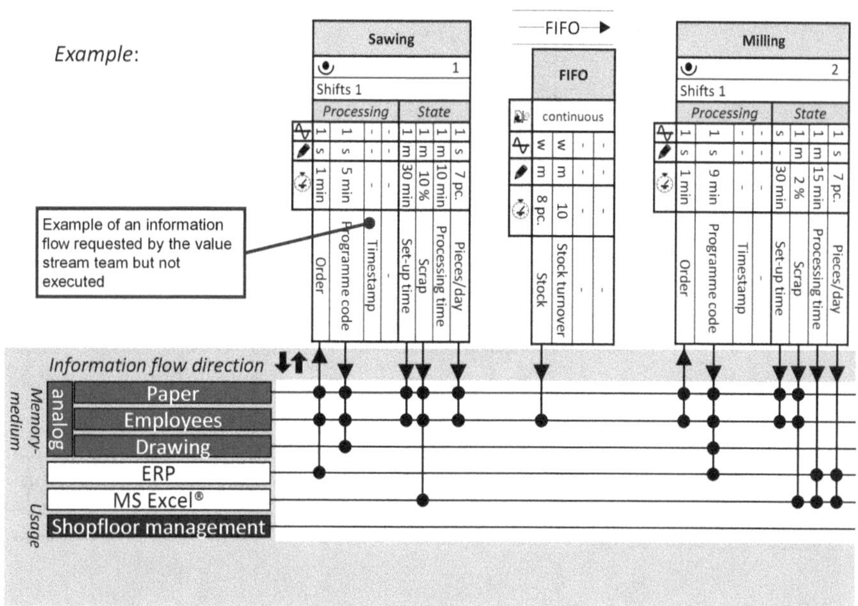

FIGURE 3.17 Value stream map up to the current step – analyzing process Information and information flows and map interconnections.

the automatic calculation of the cycle time from the log-on and log-off times or the calculation of the OEE key figure.

3.2.2.3.1 *Procedure to Map the Current State*

For each usage category (intermediate calculation or direct data usage), a separate swimlane is drawn on the value stream map, which is vertically linked to the recorded process data. However, this link is represented by dashed lines. The process owners should be asked about the use of the captured data for their own improvement activities or by other sub-processes (Figure 3.18).

3.2.2.3.2 *Guiding Questions*

- What purposes and frequency drivc KPI utilization?
- Is there an established shop floor management protocol conducted at regular intervals?
- Are improvement measures triggered based on the accumulated data?

3.2.2.4 Step 8: Capture Information Logistics Waste

Eight different categories include potential cases of information logistical waste (defined in the previous section). It is not mandatory to assign each identified waste to a specific category. Instead, these categories serve as a tool to focus on inefficient information processing. All identified information logistics inefficiencies should be documented across all process stages and logistics components.

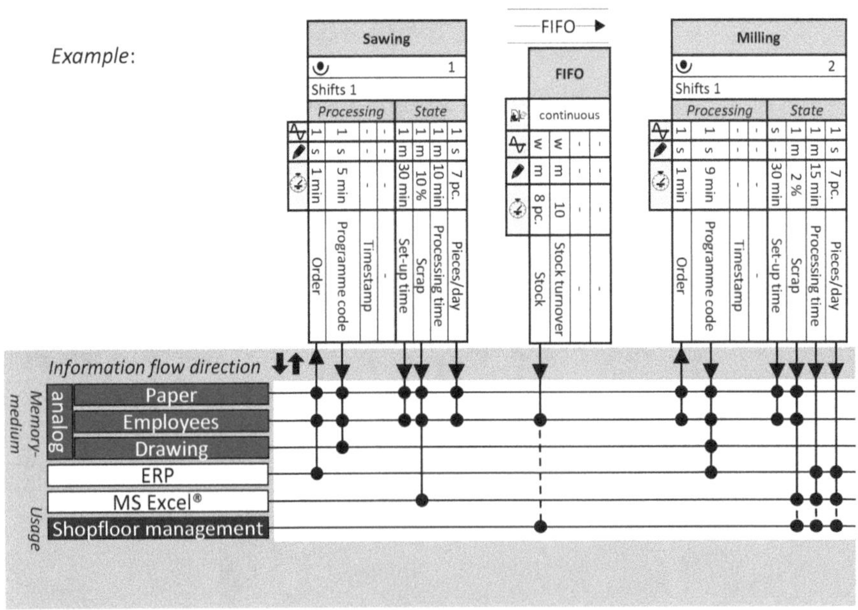

FIGURE 3.18 Value stream map up to the current step – evaluating data usage and map interconnections.

3.2.2.4.1 Procedure to Map the Current State

Thorough analysis is directed toward each activity or KPI documented within a process stage or logistics element of the value stream, assessing information handling inefficiencies. These recognized inefficiencies are itemized beneath the swimlanes, written in a dedicated section for Kaizen. This approach, differing from the traditional VSA, acknowledges that VSA4.0 identifies further forms of waste, and listing all waste types as Kaizen flashes on the value stream map can swiftly lead to confusion (Figure 3.19).

3.2.2.4.2 Guiding Questions

- The guiding questions are listed in the description of each waste category.

3.2.2.4.3 Extended Value Stream Indicators

To quantitatively characterize information flows within the value stream and to facilitate future pre- and post-improvement comparisons, aided by the implementation of Value Stream Design 4.0, three essential metrics are introduced (refer to the appendix): Digitization rate, data availability, and data utilization. These metrics can be assessed for the current state at this moment.

3.2.2.5 Step 9: Derive and Prioritize Kaizens from Information Logistic Wastes

In the final step of VSA4.0, the identified traditional and information logistic inefficiencies must be categorized into distinct improvement subjects (Kaizen initiatives).

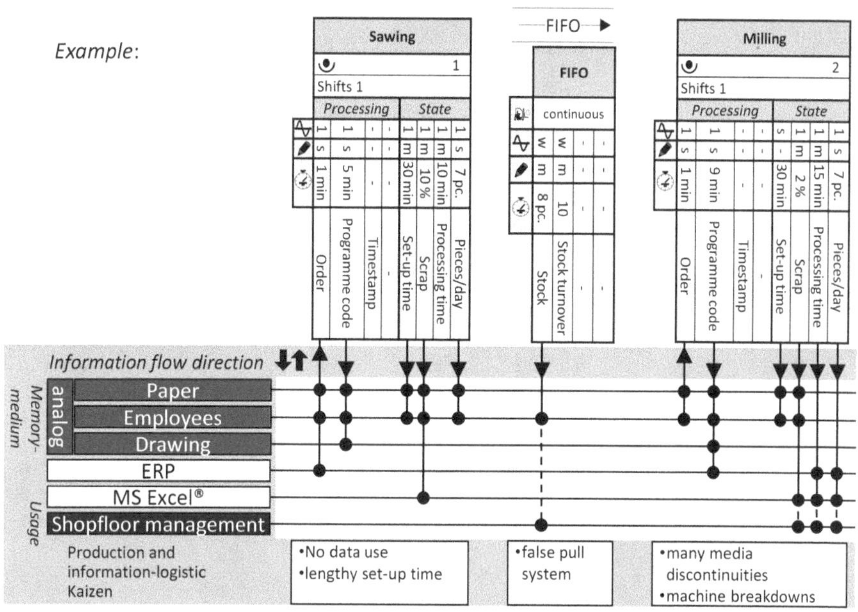

FIGURE 3.19 Value stream map up to the current step – information logistical waste added.

To perceive the significance of these Kaizen activities within a value stream, it can prove advantageous to assess individual Kaizens through a cost-benefit analysis rather than relying solely on intuition. This approach enables a focal point on Kaizen activities delivering the greatest beneficial impact, aligning with management's established priorities. Such initiatives can be effectively addressed during subsequent VSD4.0 phases. Moreover, this approach prevents addressing easily resolvable issues too early within the scope of improvement projects.

3.2.2.5.1 Optional: Prioritizing Kaizens through a Cost-Benefit Analysis

The identified Kaizens are arranged in a linear sequence (a spreadsheet program is recommended). Following this, the target dimensions such as quality, costs, or delivery reliability can be assembled and comparatively ranked. These target dimensions are inserted into columns to establish a matrix. Subsequently, the effect of implementing a Kaizen on each target dimension requires assessment (e.g., 1 = minimal impact, 5 = highest impact). By multiplying a Kaizen's benefit contribution by the weight of the respective target dimension, the total weighted benefit contribution for that Kaizen is determined. After calculating the total benefit contributions for all Kaizens, they can be organized in descending order based on their benefit value. To conclude the process, a preliminary estimate of the (financial) effort needed to address a Kaizen is included. If a subsequent engagement with Value Stream Design 4.0 is not envisaged, one can commence problem-solving for the prioritized Kaizens and address them through project work and PDCA. Nevertheless, it is prudent to complete VSD4.0 in its entirety to foster a wholly novel perspective across the entire value stream (Figure 3.20).

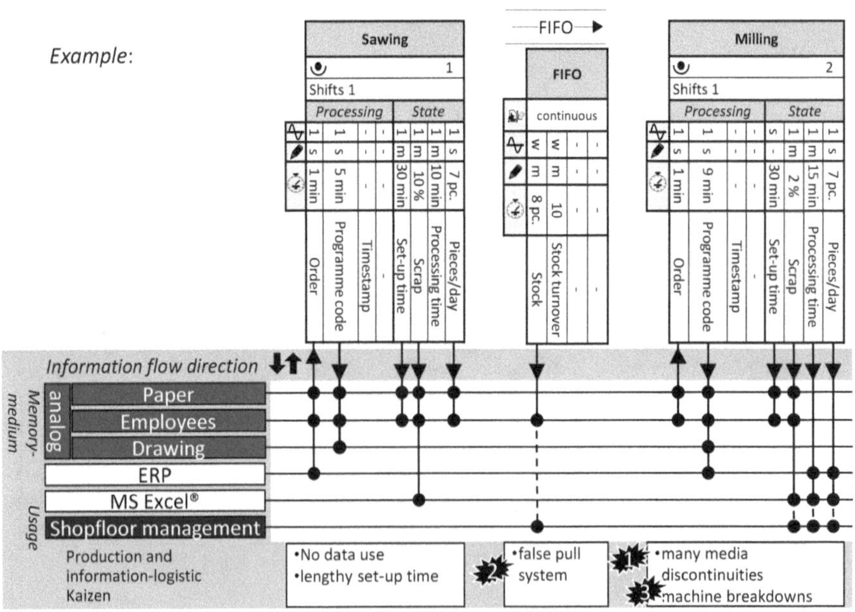

FIGURE 3.20 Value stream map up to the current step – prioritization of Kaizen and wastes.

3.2.3 Practical Example – Application of Value Stream Mapping 4.0

The Pumpen AG, manufacturer of custom connection flanges, receives specific measurements from various sources, known as storage media. The creation of individual customer orders takes place within order acceptance. This includes multiple checks of faxes, letters, and emails, as well as recording telephone orders. For the first three storage media, customers independently generate their order information. However, it appears that this information is occasionally incomplete, leading to employees requesting missing details. This results in significant additional effort, prolongs processing time, and leads to additional inquiries, especially when direct customer contact is not possible (Figure 3.21).

In the next step of the construction process, various storage media are used for the adaptation development tasks. The process reveals significant discontinuities between the systems and a predominantly manual approach. Even in CNC programming, machine codes are tailored to individual customer requirements, and despite the use of IT, manual actions still prevail, accompanied by gaps in media. For example, information is transferred via USB sticks before further processing. Remarkably, data generated in the indirect area is not systematically utilized for process improvement (e.g., shop floor management) in terms of deviations or errors.

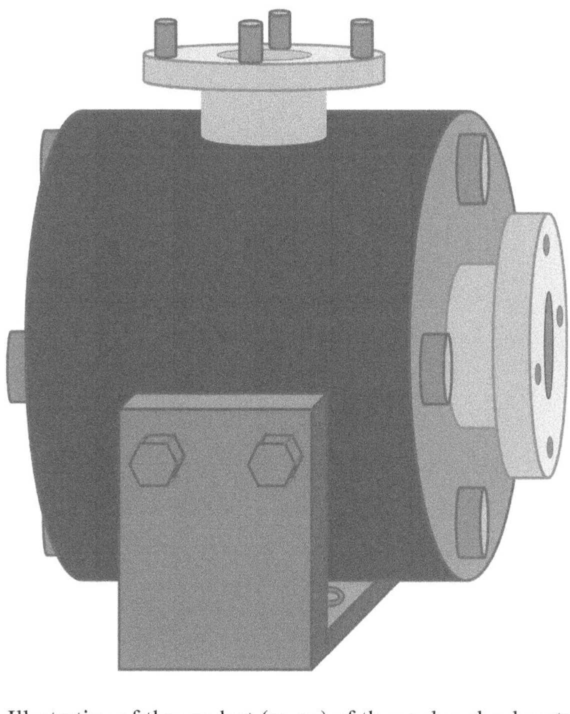

FIGURE 3.21 Illustration of the product (pump) of the analyzed value stream example of Pumpen AG.

In the direct area, Key Performance Indicators (KPIs) are logged during production and assembly and subsequently discussed in shop floor management. However, not all desired KPIs are consistently recorded and considered. Both processes heavily rely on analog information. At the end of the value stream, no KPIs are captured in the shipping process, highlighting the absence of independent creation of shipping labels from the ERP system. Employees manually transmit information to the respective shipping service providers, which is associated with significant effort and potential for errors.

The in-depth analysis of Pumpen AG's value stream was documented as part of a Value Stream Map 4.0 (see next page). Additional identified Kaizen:

- Mainly analogue storage media are used for information transport, consuming substantial processing time.
- In construction and CNC programming, repetitive tasks of a similar nature and scope are evident (Figures 3.22–3.24).

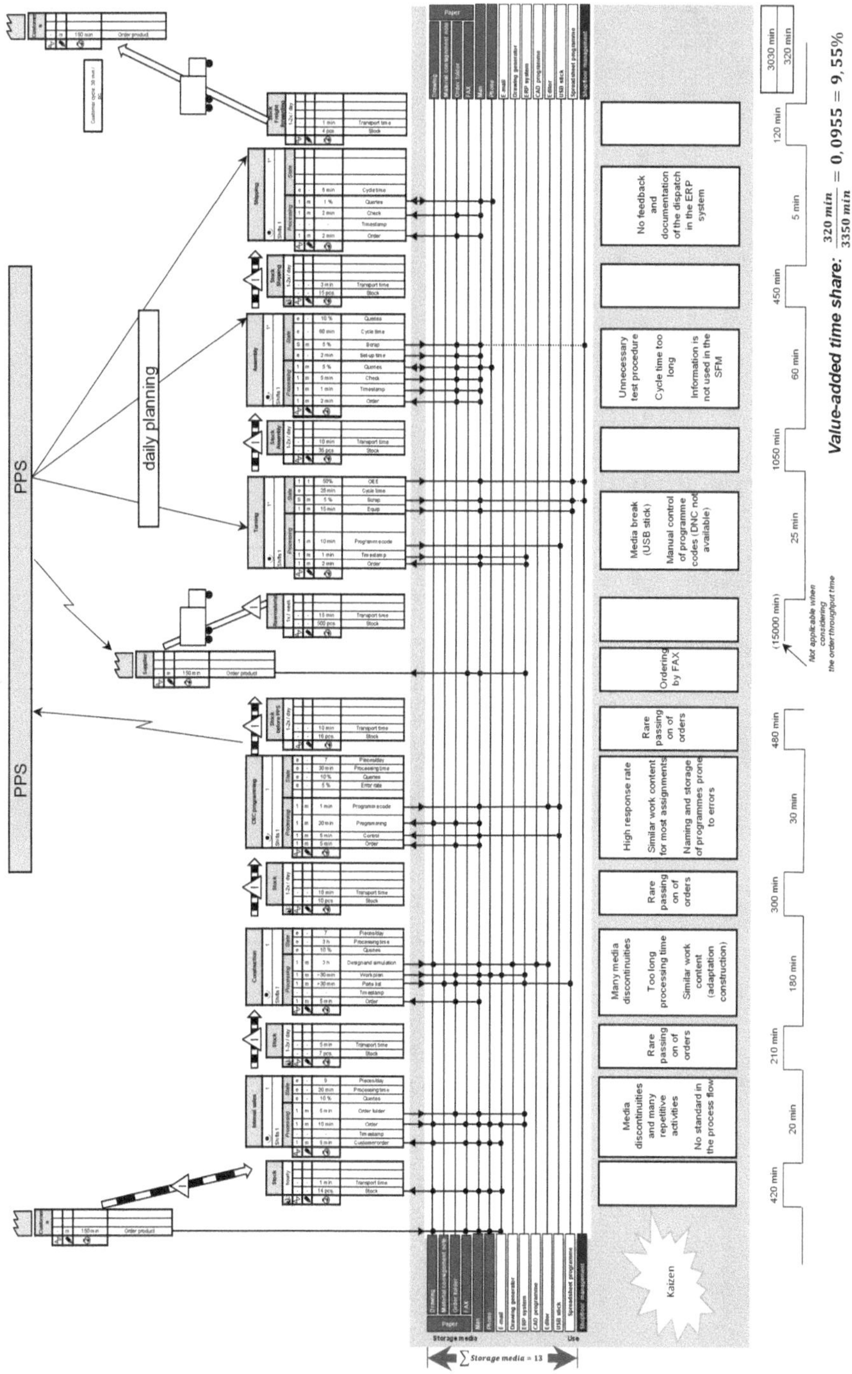

FIGURE 3.22 Value stream map example Pumpen AG – complete view.

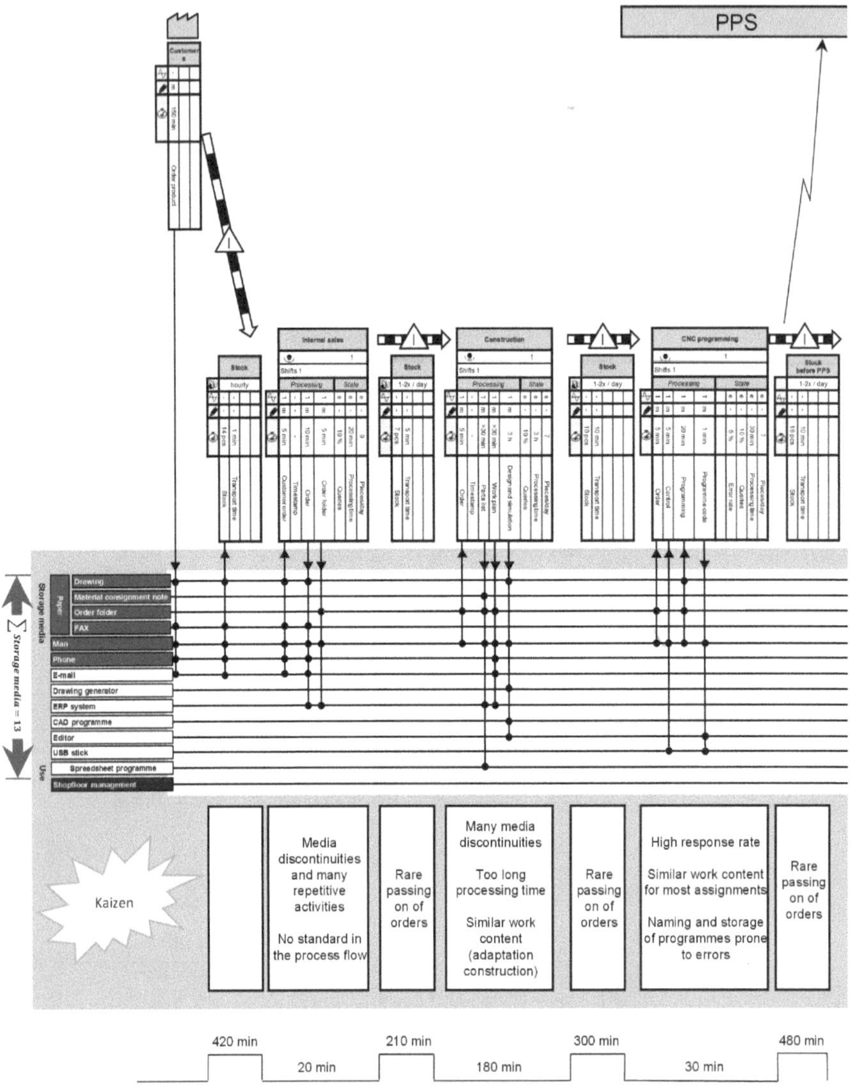

FIGURE 3.23 Value stream map example Pumpen AG – enlarged illustration 1.

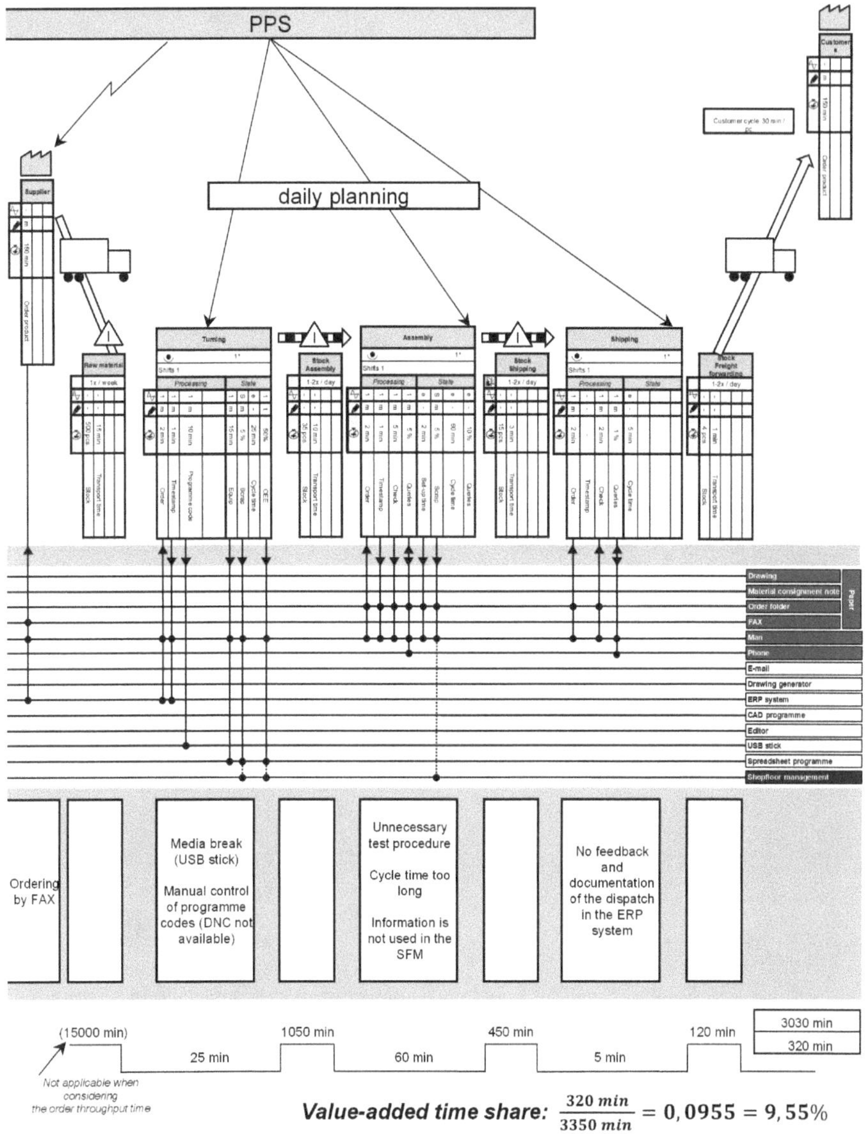

FIGURE 3.24 Value stream map example Pumpen AG – enlarged illustration 2.

3.3 WHAT HAS BEEN ACHIEVED WITH VALUE STREAM MAPPING 4.0?

See Figures 3.25 and 3.26.

Oversight of the Entire Order Processing

Value Stream Mapping 4.0 evaluates the complete order fulfilment process, assessing both the material lead time and the response time. The response time is the duration, as experienced by the customer, from placing an order to receiving the shipped product.

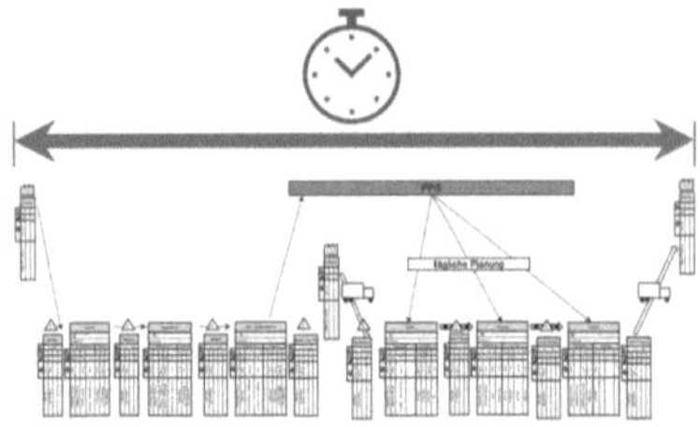

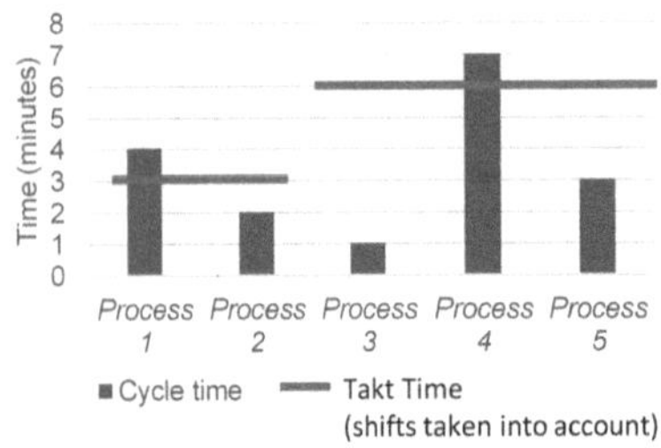

Transparency About Time Shows BottleNeck

Value Stream Mapping 4.0 enables the identification of traditional inefficiencies within a value stream. For instance, a bar chart can be employed to depict cycle time and takt time, revealing potential bottlenecks in the process.

Information Logistics Analysed

The adoption of a user-centric approach allows for the identification of inefficiencies in information processing. This approach encompasses the complete cycle of generation, transmission, handling, storage, and utilisation.

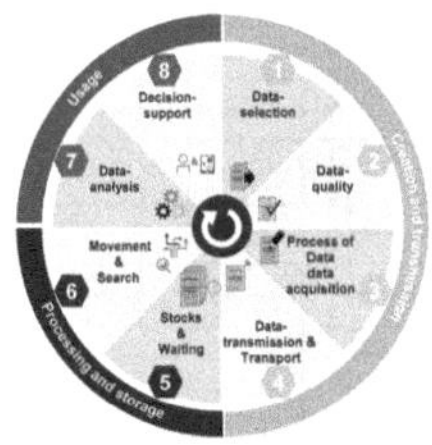

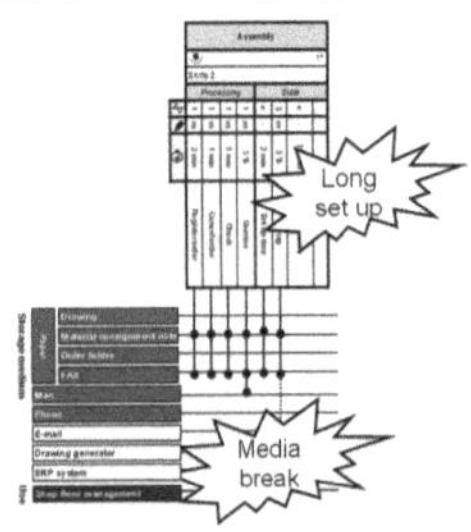

Recognition of Information Logistics and Classical Kaizens in a Unified Depiction

Through value stream mapping 4.0, it becomes feasible to scrutinise both inefficiencies in information management and the conventional 7+1 forms of wastage, depicting them as Kaizen indicators. This establishes a foundation for enhancements via lean principles and/or digitalisation.

General Overview of the Media Used

All utilised storage mediums have been documented on the value stream map 4.0. The count provides an initial insight into the intricacies of communication within the value stream. Furthermore, this aids in pinpointing redundant storage methods.

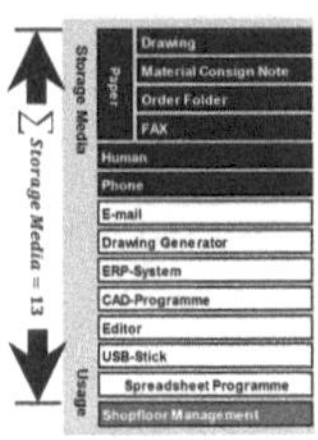

FIGURE 3.25 Value Stream Analysis 4.0 – achievement I.

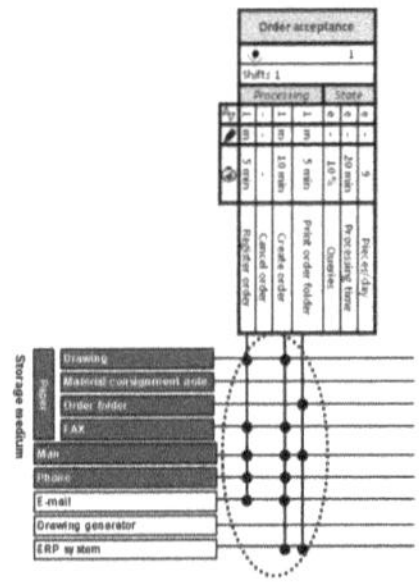

Interfaces Must be Clearly Displayed

Inefficient information processing is hindered by incompatible interfaces. Interfaces linked to paper-based or analogue mediums can be readily identified. Any issues related to digital interfaces should be investigated on location.

Digitisation Pools Uncovered

On the Value Stream Map 4.0, opportunities for ongoing digitalisation are readily apparent. It becomes evident whether processes primarily rely on digital or analogue communication. Dominant media can be identified and standardised.

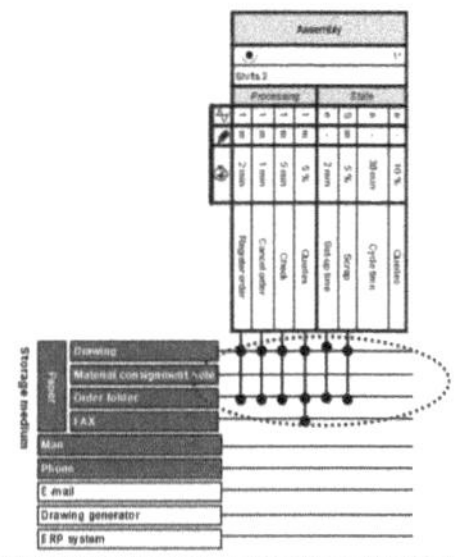

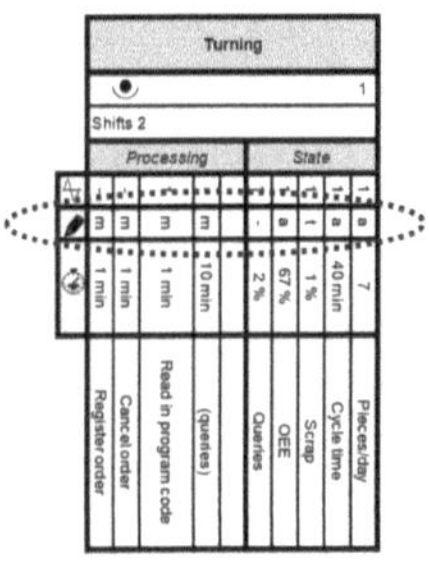

Automation in Information Information Processing Presented

The introduction of new process boxes accurately portrays the information's availability. This provides insight into whether information processing is automated (a), partially automated (t), or involves manual interaction (m). The time taken for manual process steps serves as an initial indicator for assessing the feasibility of automation.

Use of Information Visualised

If information contributes to process enhancement, this is denoted by a distinct link to an activity (dashed line). This facilitates the observation of whether the gathered data genuinely enhances transparency regarding process performance, and if identified discrepancies indeed instigate improvement efforts. Instances of this application include daily shop floor management or predictive maintenance.

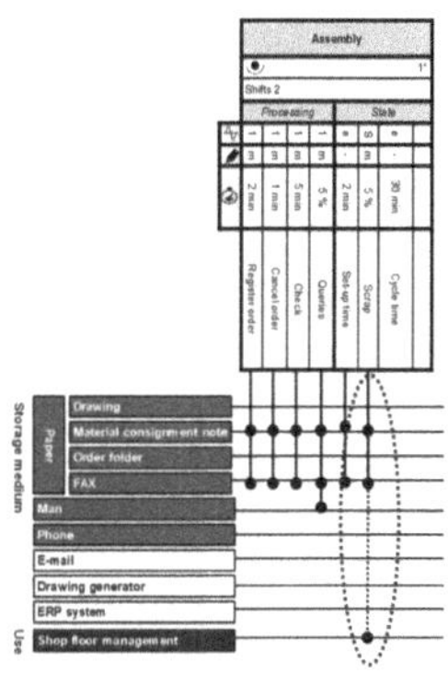

FIGURE 3.26 Value Stream Analysis 4.0 – achievement II.

NOTES

1 The classic method of value stream analysis is described in Rother/Shook (2015) as well as in Erlach (2010). Here, we have adjusted the steps to align with our experience of on-site pragmatic analysis. For instance, we have combined method steps that can be carried out simultaneously in one process or department (see Step 2 of the method described here).
2 Further analysis and key figures in Erlach (2010).
3 Meudt (2020); Meudt, Metternich, Abele (2017).
4 Cf. Dinter and Winter (2008).
5 Cf. Meudt, Leipoldt and Metternich (2016); Meudt, Metternich and Abele (2017).
6 Cf. Meudt (2020), Meudt, Metternich, Abele (2017).

REFERENCES

Dinter, Barbara; Winter, Robert (2008): *Integrierte Informationslogistik*; Berlin, Heidelberg: Springer.

Erlach, Klaus (2010): *Wertstromdesign - Der Weg zur schlanken Fabrik*; 2., edit. und erw. ed.; Berlin, Heidelberg: Springer.

Meudt, Tobias (2020): *Wertstromanalyse 4.0- Eine Methode zur integrierten Erfassung und Analyse von Material- und Informationsflüssen in Wertströmen*; Düren: Shaker Verlag.

Meudt, Tobias; Leipoldt, Christoph; Metternich, Joachim (2016): The new look at waste in the context of Industrie 4.0; *ZWF Zeitschrift für wirtschaftliche Fabrikbetrieb* 111 (11), pp. 754–758.

Meudt, Tobias; Metternich, Joachim; Abele, Eberhardt (2017): Value stream mapping 4.0-Holistic examination of value stream and information logistics in production; *CIRP Annals* 66 (1), pp. 413–416.

Rother, Mike; Shook, John (2015): *Learning to See - Increasing Value and Eliminating Waste with Value Stream Design; Version 1.4*; Mühlheim an der Ruhr: Lean Management Institute.

4 Value Stream Design 4.0 – Understanding and Guidelines

4.1 MODEL FOR SYNCHRONIZED INFORMATION AND MATERIAL FLOW

Taking inspiration from Taichi Ohno, the visionary behind the Toyota Production System, the framework of Value Stream Design 4.0 (VSD 4.0) extends its reach to integrate the "fuzzy" front end of the order fulfillment processes into Value Stream Thinking. The belonging stages of order clarification and process preparation often witness only little material movement. Here information is the primary product. In striving for the utmost efficiency in customer order processing and to foster a seamless flow of customer orders, attention must be devoted to the interplay between preparatory activities, auxiliary functions, and production itself. To facilitate this, we extend the notion of a product, delineating between tangible material and intangible information. In the auxiliary domain (e.g., adaptation engineering, work preparation, etc.), the product initially comprises solely data and information, warranting our focus on the "order processing information flow" at this juncture. In the direct sphere of production, the product flow involves physical materials or tangible products. We examine this within the ambit of the "product material flow". It's the orchestrated convergence of these two flows that culminates in the most efficient reduction of throughput times.

The **flow of order processing information** encompasses all necessary data essential for processes to effectively execute their functions and duties or is generated within their purview. Its origination starts with the customer and progresses through the intermediate process steps. In these processes, information undergoes further refinement and preparation, facilitating the production of the physical product within the manufacturing realm, to meet the specifications of the respective order. Illustratively, the information processing flow incorporates e.g. parts lists, order specifications, and test plans.

The **flow of product material** embodies the movement of physical materials within the direct operational processes. In the realm of production, this flow converges with the flow of order processing information.

Both flows together form the **product flow**. This, in turn, is bolstered by an additional duo of flows: the flow of status information and the flow of auxiliary resources.

The **flow of status information** encompasses overarching details regarding the state, utilization, or effectiveness of the processes (including factors like processing

DOI: 10.1201/9781003586906-4

time, temperature, resource consumption, pressure, etc.). This status information flow finds application in various scenarios, such as fostering visibility into production, serving as a foundation for process enhancements, documentation, or facilitating component traceability.

The **flow of auxiliary resources** encompasses the supply and conveyance of essential operational and supporting materials required for order execution, such as in machining or assembly. It falls outside the purview of VSD 4.0, as it constitutes a distinct value stream necessitating analysis within a separate project.

The effective configuration of each of the four flows exerts an impact on the overall efficiency of the value stream. Notably, the alignment and coordination of the order processing information flow and the product material flow must be emphasized (Figure 4.1).[1]

If any of the four flows encounters difficulties or falls out of sync with the others, delays arise due to waiting periods. To circumvent this issue, it is imperative to establish an integrated design for these information and material flows. This is notably complex within the realm of production, where all four flows converge: at a production workstation, concurrent access to the associated work and inspection instructions must align with the physical product. Simultaneously, tools, equipment, and measuring devices must be readily available and the belonging process parameters must be configured. Furthermore, the customer should be linked to the value stream as a source of information and the initiation of order processing, enabling efficient organization of order clarification, adaptation engineering, and work preparation.

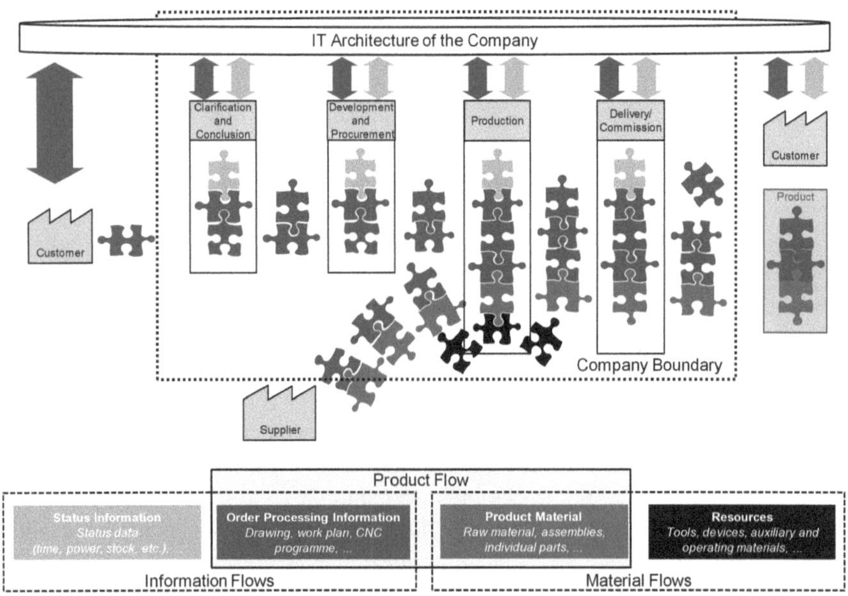

FIGURE 4.1 The different parts of the value stream. (*Source*: Hartmann 2022; Metternich 2018.)

4.2 GUIDELINES FOR DESIGNING LEAN INFORMATION FLOWS

In line with the principles of creating a streamlined material flow, the following guidelines aim to offer direction for creating efficient information flows.

Only Digitize Lean Processes

Lean precedes digital! The optimal synergy of material and information flows is most impactful when waste has already been eradicated, work procedures are well established and consistent and material flows smoothly. Introducing digital solutions into inefficient processes risks digitally supported standardization of waste. Consequently, it becomes challenging to pinpoint and eliminate the underlying issues. Therefore, it is advisable to fundamentally eradicate identified waste before considering digital solutions. A beneficial approach is to apply the "5 Whys" method to every identified waste (termed "Kaizen flash") to surface the root cause. This root cause should primarily be addressed using traditional lean methods. Digital options should only be explored if these prove insufficient.

Get the Information Flowing

Unlike the flow of materials, delays in the flow of information are less apparent. The seamless information flow gets interrupted when orders cannot progress due to missing data or the need for questions. It's crucial to methodically identify the information required by a customer or subsequent process to enable the immediate processing of an order. Ensure this information is readily available or that existing data is consistently and promptly updated through suitable data connections. Pay particular attention to paper-based information, which carries the risk of rapid obsolescence (e.g., work and test instructions, packing lists, parts lists, etc.). In time and quality critical order processing, maintaining a continuous information flow among all value-adding processes (mechanical production, pre and final assembly, inspection, and testing) is essential to prevent delays caused by transportation times.

Standardize and Automate Information Processing

Standards facilitate the stabilization of desired process performance. On the one hand, they involve adhering to the current best practices, while on the other hand, they allow identifying and addressing deviations from the expected standard performance. This principle applies to both traditional forms of waste and information logistics waste. Initially, it is crucial to establish stable and standardized processes. These processes subsequently streamline the adoption of digital and automated solutions. It's beneficial to consider the viewpoint of information users: who needs specific information, in what format, and at what time? Furthermore, it's essential to identify the information sources. For each piece of information, establish a consistent and uniform information carrier, along with standardized methods for collection (e.g., input forms or scanning devices), transmission, presentation, and utilization, typically within the framework of standardized communication protocols.

Meet Complexity with Simplicity

Enhance established lean methods through digital integration rather than substituting them entirely with digital solutions. Digital systems seldom provide the sole solution to a problem. Begin by refining lean processes through the elimination of waste and the root causes of instability. Subsequently, a digital system can enhance the identified solution. For instance, technologies like pick-by systems or intelligent data glasses reach their full potential when the picking sequence and material arrangement are pre-coordinated in a picking area.

Do Not Allow Information Flow without Information Use

Every information flow should have a recipient who actively uses the information. This recipient can be another process step or a support process, such as maintenance or logistics. Utilization can also involve data analysis, which aids shop floor management or enables process monitoring. Information flows that lack utilization are wasteful.

Use the Potential in Information

When employed effectively, data can offer visibility into the current production status and support improved decision-making. Unutilized measurement records or test bench outcomes represent a squandering of valuable insights into enhancing products and processes. Recognize areas where data is underutilized, such as in measuring processes or test benches, and connect machine and system process data with product test or measurement results using timestamps. In situations involving quality issues, this approach can expedite problem identification and resolution.

Put the Employee at the Center of Digital Approaches

The essence of the lean philosophy centers on employees who consistently enhance processes and augment their own problem-solving capabilities. Consequently, when conceiving and rolling out digital solutions, place emphasis on employees as problem solvers. Identify areas where decisions can be enhanced through the aid of (digital) information provision. Determine how problem-solving processes can be bolstered with well-prepared data. Evaluate which digital end devices and input interfaces notably enhance employee acceptance. Also, ascertain the requisite qualifications for proficiently handling a new digital solution and devise methods for imparting them effectively (Figure 4.2).

In summary

Guidelines for your Value Stream Design 4.0

- ✓ Only digitize lean processes
- ✓ Get the information flowing
- ✓ Standardize and automate information processing
- ✓ Meet complexity with simplicity
- ✓ Do not allow information flow without information use
- ✓ Use the potential in information
- ✓ Put the employee at the center of digital approaches

FIGURE 4.2 Summary of guidelines for the VSD 4.0.

NOTE

1 Cf. on the definition and subdivision of information and material flows Hartmann (2022), Metternich (2018).

REFERENCES

Hartmann, Lukas (2022): *Wertstromdesign 4.0- Methode zur integrierten Gestaltung von Material- und Informationsflüssen für schlanke Wertströme*; Düren: Shaker Verlag.

Metternich, Joachim; Meudt, Tobias; Hartmann, Lukas (2018): Execution of the Value Stream Analysis 4.0. In *Leitfaden Industrie 4.0 trifft Lean - Wertschöpfung ganzheitlich steigern.* (Ed. by VDMA; FKM; PTW; Hartmut Rauen, Christian Mosch, Felix Prumbohm); Frankfurt: VDMA Verlag.

5 Value Stream Design 4.0 – Determining the TARGET State

5.1 PHASE III: DEFINE REQUIREMENTS FOR THE VALUE STREAM

This phase seeks to establish prerequisites for the configuration and direction of the forthcoming value stream. These prerequisites stem from both the business model and the functional domains. Figure 5.1 provides an overview of the following three phases of value stream design starting with the definition of requirements. By scrutinizing the business model, the value stream's design should be reoriented more significantly toward the creation of customer value (Figure 5.2).

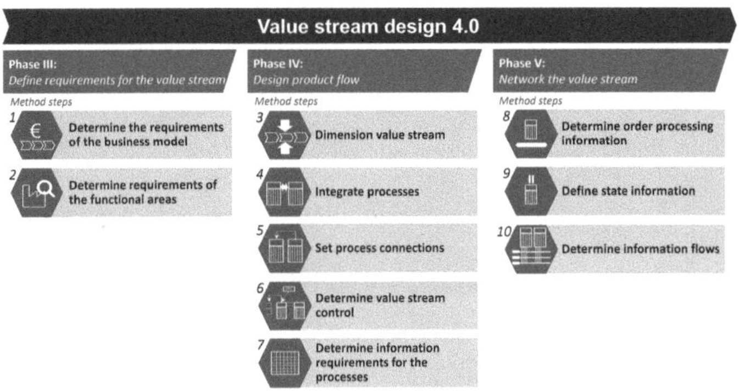

FIGURE 5.1 Value Stream Design 4.0 – method overview.

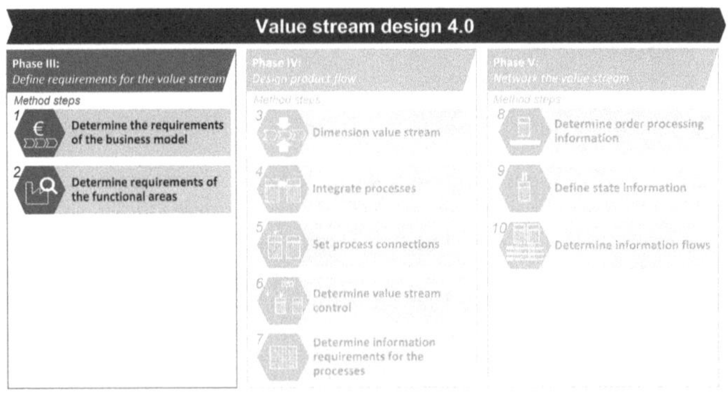

FIGURE 5.2 Value Stream Design 4.0 – Phase III.

DOI: 10.1201/9781003586906-5

5.1.1 Step 1: Determine the Requirements of the Business Model

The business model serves as the initial point for shaping the value stream. It can be broadly defined as follows:

> A business model encapsulates the fundamental principles governing how an organization generates value. This model delineates:
>
> 1. What the organization provides that holds value for customers.
> 2. How value is engendered within the organizational framework.
> 3. How these created values are conveyed and delivered to the customer.
> 4. How the value generated is accumulated as earnings by the company.
> 5. How these values are disseminated within the organization and to stakeholders.
> 6. How the foundational principles of value creation evolve to guarantee the sustainability of the business model in the future.[1]

A systematic and structured examination of a business model unveils how a company's value proposition materializes through the synergy of internal processes and strengths, in conjunction with the engagement of external partners and channels. The conventional methods of illustrating business models, primarily due to their excessive abstraction, are initially unsuitable for the specific detailing of processes. Within the framework of Value Stream Design 4.0, the task at hand is to meticulously convert the insights gleaned from the business model analysis into precise specifications for the comprehensive design of both material and information flows. This is because the value stream should portray the enactment of the business model concerning the operational provision of services. Before the actual design of the future value stream, it is imperative to ascertain the requirements that the business model imposes on the value stream.

To establish a value stream capable of delivering novel and distinctive value propositions, it is essential to address the following two questions:

> What are the current and future demands imposed by the business model on the value stream and its network of information logistics?
>
> Is it possible to harness information from the value stream to bolster the business model or generate additional value?

To precisely identify these requirements and value propositions, various approaches can be employed. A frequently adopted method, and the one utilized here, involves the use of the Business Model Canvas (BMC). It furnishes a structured design framework for delineating the business model along with its pertinent facets. The business model within the BMC is categorically divided into various segments, as depicted in Figure 5.3.

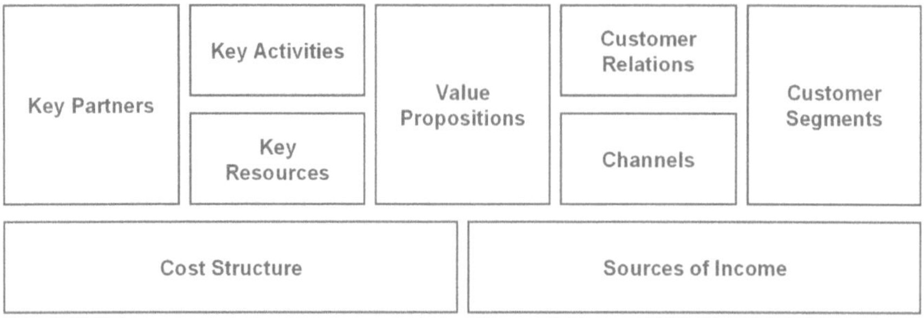

FIGURE 5.3 Business Model Canvas (Osterwalder and Pigneur 2011).

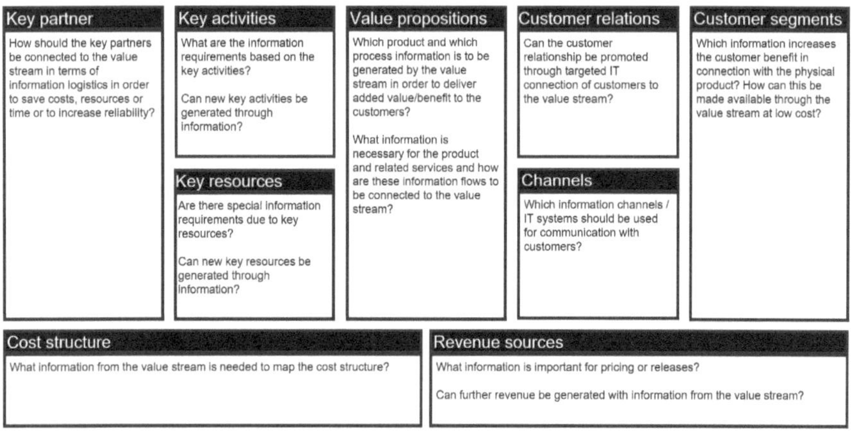

FIGURE 5.4 Business Model Canvas with guiding questions (Osterwalder and Pigneur 2011).

The formulation of the business model, encompassing its fundamental components, offers the value stream team a guiding framework for the continuous reconfiguration of the value stream. Every process should be empowered to meet the distinct requirements of the business model and its value proposition for each individual customer order. This pertains to both the material flow and the information flow.

On one hand, the business model engenders requisites for the material flow, which are to be ascertained by the value stream team and duly incorporated into the process design during Phase IV. On the other hand, the business model might impose demands on the formulation of information logistics. These demands are to be delineated for Value Stream Design 4.0. Targeted questions can aid in identifying these requisites within each element of the BMC. To facilitate this, illustrative questions are provided in Figure 5.4.

Drawing from the elements of the business model and the answers to the guiding questions, information requirements are formulated, which the value stream must meet to enact the business model. Sample information requisites are detailed below:

- Supermarket inventory data: This serves to connect suppliers to the value stream, thus streamlining the ordering of intermediate products and purchased parts, either through automation or enhanced efficiency.
- Resource and energy consumption: To align with the corporate strategy emphasizing transparency in resource usage during product manufacturing, pertinent information derived from value stream processes should be documented and provided to customers.
- Measurement and test data: As an added value proposition, offering customers access to measurement and test data from the value stream before product delivery. This can accelerate commissioning.
- Time stamps: Customers should receive progress updates on order fulfilment through the company's web portal. To facilitate this, time stamps for relevant orders and processes need to be recorded and made accessible.

These identified information needs can be incorporated into an information matrix for subsequent stages of the process, elaborated upon in the following step.

5.1.2 Step 2: Determine the Requirements of the Functional Areas

In addition to the requisites stemming from the business model, internal functional sectors or supporting processes may also present prerequisites for the value stream. Examples include quality management, order planning, and maintenance. These entities require information from the value stream to accomplish their distinct tasks and objectives. Timely access to information from the value stream can additionally enhance the efficacy of internal functional areas and streamline collaborative efforts (Figure 5.5).

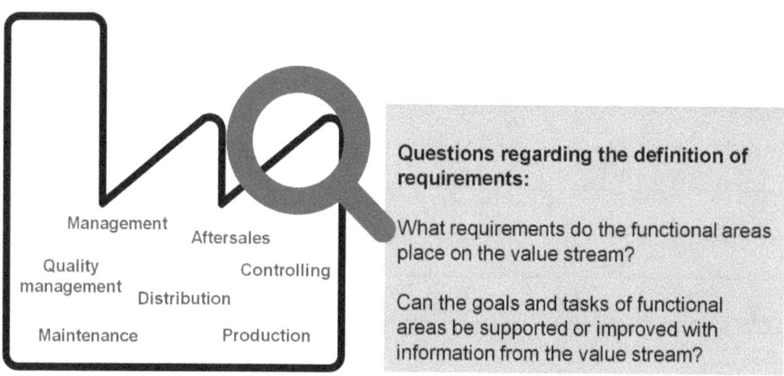

FIGURE 5.5 Assessment of requirements of functional areas.

Examples of information requirements for individual functional sectors encompass:

- Shopfloor management: quantity of units, failures, rework, overall equipment efficiency (OEE),
- Quality management: conforming parts, rejections,
- Logistics: inventory levels, storage locations,
- Maintenance: temperature, coolant levels, tool wear.

The identified information requisites, stemming both from the business model and the functional sectors, feed into the subsequent phases of Value Stream Design 4.0. In the course of the structured analysis of the business model and the functional areas, a multitude of specific information requirements are typically delineated. Hence, it can be beneficial to employ a tool within the method. To this end, we introduce an information matrix. This matrix enables the recording of information needs and their subsequent integration into the design of processes and the overall value stream. The matrix is structured as follows:

- The initial column enumerates the elements from the business model and the functional sectors that necessitate information from the value stream.
- The second column indicates the corresponding information requirements.
- In the subsequent columns, the processes within the value stream are linked to the information requirements. These processes serve as potential suppliers of the information demand of the functional areas. This section will be completed in the ensuing stages of the method (Figure 5.6).

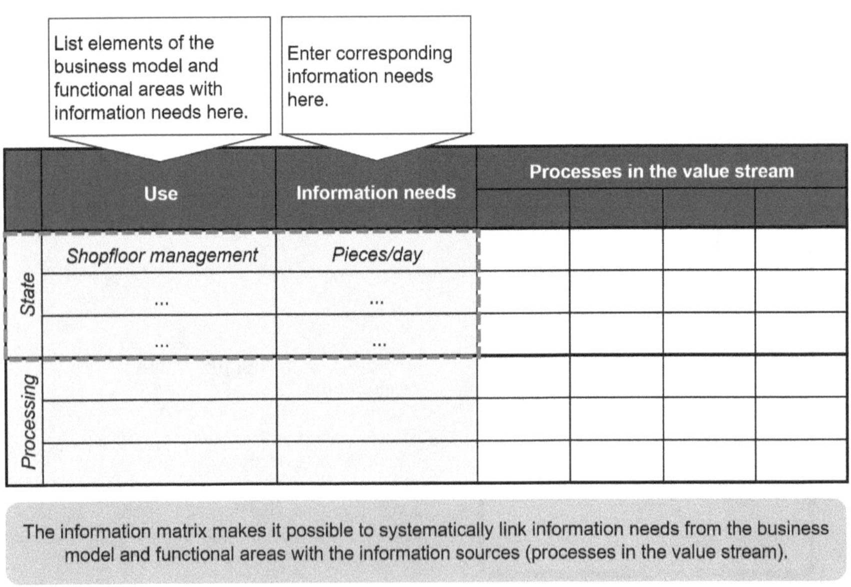

FIGURE 5.6 Structured assessment of information needs for every business model element.

5.2 PHASE IV: SHAPING THE PRODUCT FLOW

Efficient and low-waste processes, as conceived in traditional Value Stream Design, provide the foundation for a purposeful deployment of digital solutions and the synchronized configuration of material and information flows. Consequently, the subsequent steps aim to initially craft an efficient product flow following the principles of traditional value stream design. This approach aligns with the directives outlined by ROTHER AND SHOOK, supplemented with occasional decision-making aids (Figure 5.7).

The principles guiding the design of lean processes are applicable both in the direct and indirect sectors. However, it's important to acknowledge that in the indirect sector, the "product" to be produced is information. Identifying and tracking the "flow" of this product through the processes can often be more challenging than monitoring a tangible product in the direct sector. Nevertheless, the goal of Value Stream Design 4.0 is to establish processes for generating the informational product in the indirect sector (order processing information flow) in a manner akin to the processes involved in creating the material product in the direct sector (product material flow). This approach aims to ensure a coherent process design for product manufacturing. In essence, the processes in the indirect and direct sectors differ in the following criteria:[2]

- Product: In the indirect area, the product is essentially information. Information, by its very nature, is intangible and imperceptible, often necessitating a medium like paper to render it visible.
- Process transparency: Given the intangible nature of information, achieving transparency regarding processes, activities, and value stream identification is generally more challenging in the indirect sector.
- Process standardization: Processes within the indirect area tend to have lower levels of documentation and standardization compared to those in the direct sector.

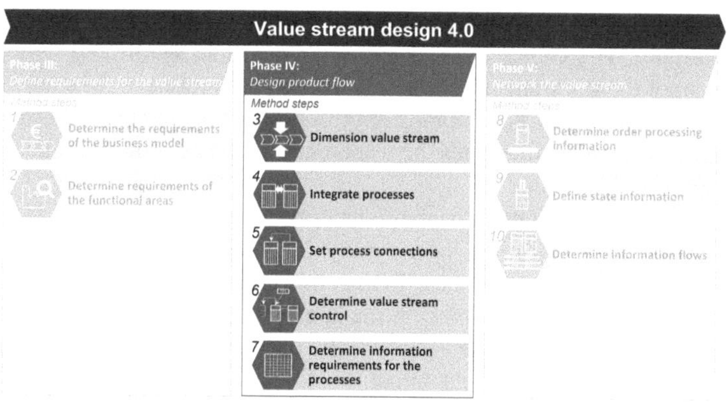

FIGURE 5.7 Value Stream Design 4.0 – Phase IV.

- Diversity and division of tasks: In the indirect sector, employees typically engage in various functions and processes, resulting in lower repetition frequency of activities and a larger number of individuals involved in the process compared to the direct sector.
- Process duration: Processing times for different tasks often exhibit more significant variations in the indirect sector than in the direct sector.
- Interfaces: Processes in the indirect area typically entail a higher number of interfaces than those in the direct sector. This can be attributed, for instance, to more substantial integration into diverse service provision processes or a larger internal customer base for these processes.

5.2.1 Step 3: Dimension the Value Stream

Before embarking on the actual process design, it is imperative to define the specific target variables and the contextual parameters for the value stream. These elements offer valuable insights for the capacity-driven design of the processes.

5.2.1.1 Establish Performance Objectives for the Value Stream

In Chapter 2, as part of preparing for a successful project, it was recommended to "quantify project objectives using key performance indicators within the processes". In line with this, extract the performance metrics outlined in Section 2.3 and articulate specific target values to be attained through the value stream design.

5.2.1.2 Defining the Principle for Order Processing

Before delineating or structuring the sub-processes within the value stream, it is imperative to establish a foundational understanding of the fundamental approach that should underpin order processing. Based on this understanding determine the principle of order processing. The basis for making this determination is the business model and its associated performance criteria. The various order processing methods have a substantial impact on aspects such as flexibility, delivery timelines, inventory management, and customer service. These methods can be broadly categorized as make-to-stock, assemble-to-order, make-to-order, engineer-to-order, and develop-to-order. The choice of the appropriate principle significantly impacts the management of the value stream, the interconnection of processes (whether it's FIFO or supermarket pull), and the shipping approach (direct shipping or finished goods warehousing). Figure 5.8 illustrates the five order processing principles and their corresponding control domains.

If a customized order runs through production, the individual stations of the value stream must be adjusted to this individuality. This applies to materials, programs, tools, devices, and work instructions. To map this individuality, production planning must provide the relevant stations with the appropriate order information. The earlier individual orders arise in the value stream, the earlier the individual scheduling of orders starts. This point of scheduling divides the value stream into a customer-neutral and a customer-individual part.

It is important to note: The earlier individual variants are created, the

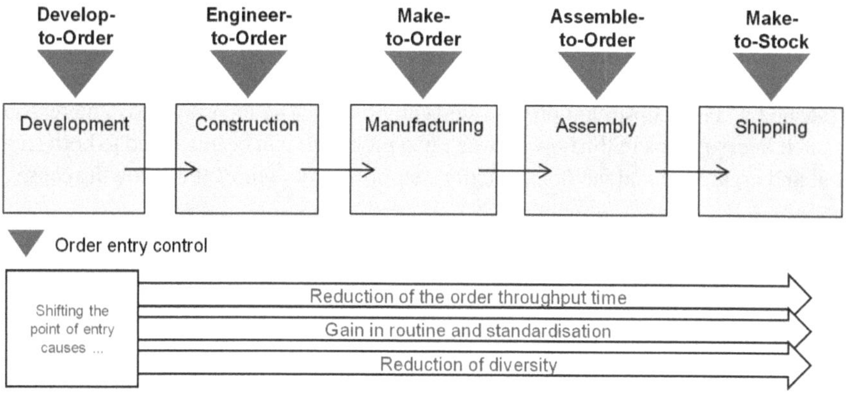

FIGURE 5.8 Order entry point and production control.

- longer a distinction must be made between these variants in the value stream;
- more complex material provisioning becomes;
- more frequent setup is required;
- more opportunities there are for errors;
- more effort takes production planning.

Nevertheless, many companies incorporate individualized solutions as a core aspect of their value proposition. To minimize the impact of this individuality on the value stream, it's advisable to limit the depth of customer order penetration and delay the manufacturing of variants, opting instead for assembly at a later stage. This requires the development of efficient modular product structures, where optional functions are isolated in mountable functional modules. The design flexibility of the value stream greatly hinges on the product structure. If some components may require an engineer-to-order approach, the focus should be on reducing adaptation efforts through close customer engagement in the digital front end and maximizing automation in adaptation development and work preparation (WP). Seamless digital continuity, spanning from requirements description to feasibility checks, CAD modelling, NC program generation, as well as work and test instructions, is a prerequisite for this.

5.2.1.3 Determining the Customer Takt

The customer takt serves as the central reference point for the capacity planning of processes and the value stream in total. Its calculation and significance are detailed in the analysis section, and thus, reference is made to it here (Section 3.1.2). When calculating the customer takt, it's essential to consider future demand requirements that the new value stream will need to fulfil.

5.2.2 Step 4: Integrate Processes

Commencing from the final process within the value stream, the one closest to the customer, the objective is to establish closely connected upstream areas in the value stream where each order progresses as an individual piece or single set without delay.

The concept of "pulling" from supermarkets should exclusively apply where this continuous flow becomes impractical ("Flow where you can, pull where you must!"). We refer to these seamlessly integrated areas as "Islands of Flow". The end of a flow island is reached when an upstream process step can no longer be incorporated. The reasons for such interruptions in the flow between two processes can be attributed to both technical and organizational factors, typically stemming from one of the following causes:[3]

- Physical separation between processes (can they be spatially consolidated?);
- The merge or division of value streams (can "racer lines" be established?);
- Unstable processes, often due to quality issues, missing material or frequent inquiries (evaluate Kaizen activities!);
- Batch formation driven by technological constraints (explore alternative technologies!);
- Varied shift schedules (can activities be synchronized within the same shift?);
- Irregular cycle times among successive processes (investigate alternative technologies and/or Kaizen to equalize cycle times!);
- Extended setup times (consider quick setup initiatives!).

Figure 5.9 illustrates an industrial scenario with four process steps that are merged into a single process marked by continuous flow (outlined by the dashed line within the diagram). By creating this "island of flow" within the value stream, the stock between process steps is diminished, planning is simplified and lead times reduced. This outcome is made possible, in this instance, due to the close similarity in cycle and setup times.

5.2.2.1 Eliminating Waste to Create Islands of Flow

The foundation for interlinking process steps lies in the presence of lean, stable, and standardized processes. Therefore, the process steps or flow areas should undergo an analysis to identify and eliminate waste and improve stability. Waste can manifest itself in both classical and information logistic categories. Recognized waste (and associated "Kaizen" opportunities) should be substantially reduced before implementing a flow island.

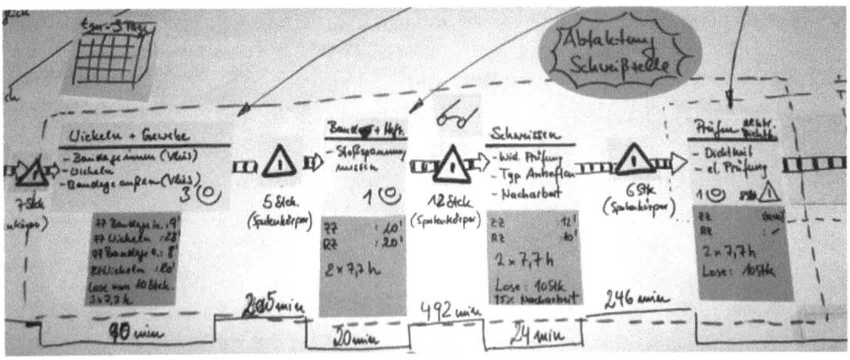

FIGURE 5.9 Creating islands of flow – practical example.

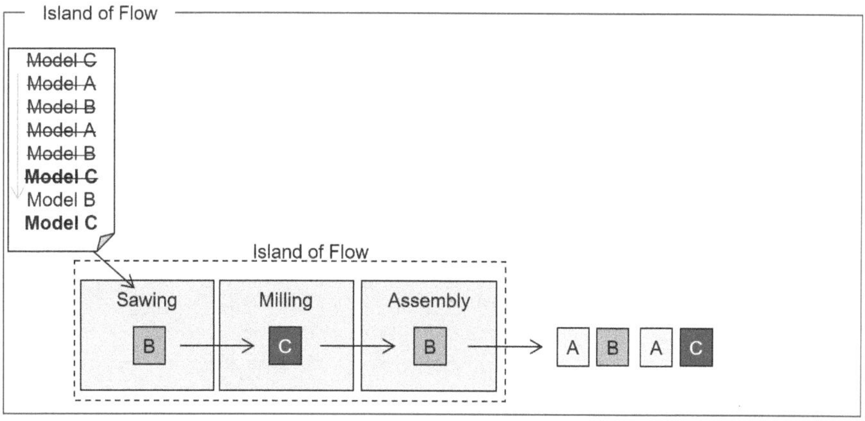

FIGURE 5.10 Island of flow – principle representation.

The waste identified during Value Stream Mapping 4.0 should be prioritized and addressed as part of the project work. To prioritize Kaizen activities, the primary objective should be the creation of an uninterrupted order flow. The key question is: What impedes the direct flow of an order from process A to process B? Below are some example improvement projects that arise from this question:

- Identify the information needs of processes and provide necessary information in a standardized and timely manner to prevent inquiries in order processing.
- Analyze the root causes of inquiries and address them.
- Resolve quality issues to reduce safety buffers and rework.
- Implement a standardized material replenishment process to improve material availability.
- Clearly separate direct line work from material replenishment tasks and assign replenishment tasks to a specialized function, e.g. a "line feeder".
- Balance workloads between stations to achieve a synchronized flow of individual pieces.
- Reduce, stabilize, and standardize changeover times of interconnected processes.

This consideration should encompass both manual tasks and the time allocated for handling information acquisition or processing (Figure 5.10).

WHY SINGLE-PIECE FLOW, ANYWAY?

The objective of single-piece flow production is often stated, but the underlying reasons are frequently not adequately elucidated, leaving management and employees perplexed. It is imperative to underscore that single-piece flow is not the ultimate goal of process improvement but rather a methodology that facilitates the attainment of the subsequent objectives. Striving for single-piece flow we achieve:

- Minimum space requirement (achieved by arranging workstations at the closest feasible proximity).
- Enhanced transparency (e.g., facilitating target/actual comparisons for output or errors).
- Optimal conditions for standard work (maintaining an even workload within the customer takt, fostering positive routines).
- Reduced inventories (minimal or no buffers between stations).
- Minimal lead time (accomplished through low inventories and the use of FiFo).
- Streamlined workstations (holding only essential tools and aids in stock, thereby reducing search times and costs).

However, the most pivotal achievement, from both a quality and improvement standpoint, is the rapid detection of defects facilitated by single-piece flow. This is because there is no mass transfer; each part is inspected and processed immediately. Consequently, errors and defects are identified promptly, permitting an immediate stop in production and the initiation of process improvements. Moreover, this approach fosters a learning environment for the team.

5.2.3 Step 5: Set Process Connections

In this phase, the formed flow islands are to be interconnected using pull systems. Pull systems establish a self-regulating connection, where the internal customer process signals a demand to the upstream process, ensuring that only the correct and currently required quantity is produced. There are various pull systems to consider:

5.2.3.1 FiFo System (First-In, First-Out)

With FiFo-Pull, the production order sequence is determined by the order in which materials arrive from the upstream process. The permission to produce (as a pull signal) originates from the downstream process. The sequence of controlled orders remains unchanged in a FiFo coupling. Pull signals can take various forms, such as an available parking space, a paper card, or a message within an IT system. Another feature of FiFo is the limited buffer size. If the FiFo buffer is full, the supplying process stops its production (Figure 5.11).

5.2.3.2 Supermarket Systems

In the supermarket system, both the quantity and type of information originate from the downstream (customer) process. Consequently, it needs to be managed through a production plan or another pull system. To limit the amount of inventory supermarkets find application in connecting processes if the demand of the downstream

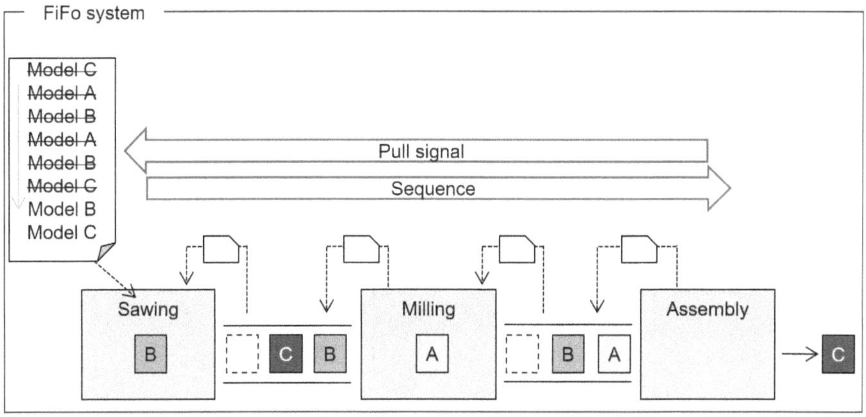

FIGURE 5.11 FiFo system.

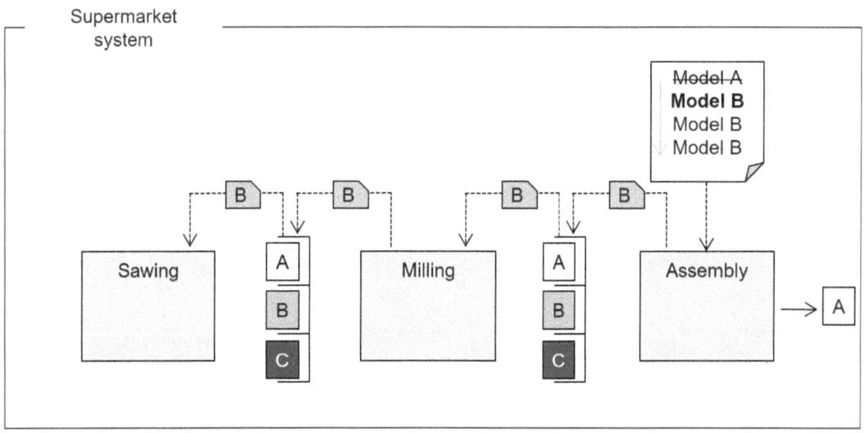

FIGURE 5.12 Supermarket system.

process shows little variation in mix and volume of units. They are also frequently employed for providing production with auxiliary materials or C-parts (Figure 5.12).

5.2.3.3 Mixed-Pull Systems

In this setup, FiFo and supermarket systems are working in parallel: commonly produced or standard items are regulated using a supermarket system. The production of these items is controlled based on rotating Kanban and released through withdrawal by the downstream process. In contrast unique and order specific items are managed through a FiFo lane. A pull signal by the downstream process triggers replenishment while a production plan tells what to replenish next (Figure 5.13).

The suitability of the systems presented in relation to specific criteria is outlined in the following overview, offering guidance for the selection of process connections (Figure 5.14).

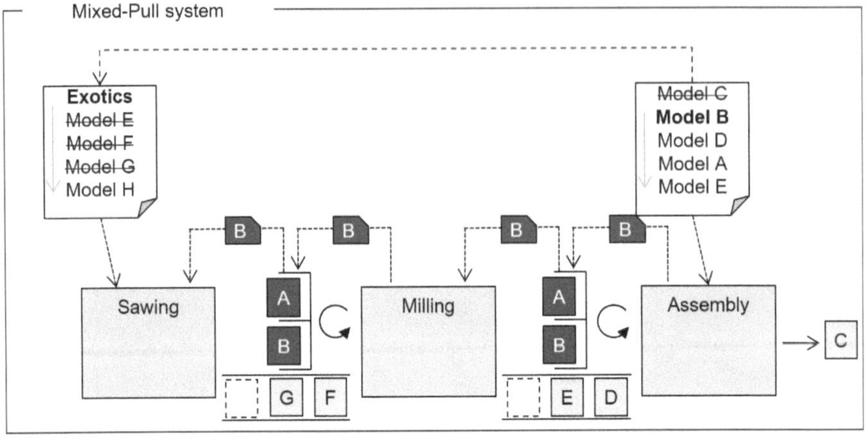

FIGURE 5.13 Mixed-Pull system.

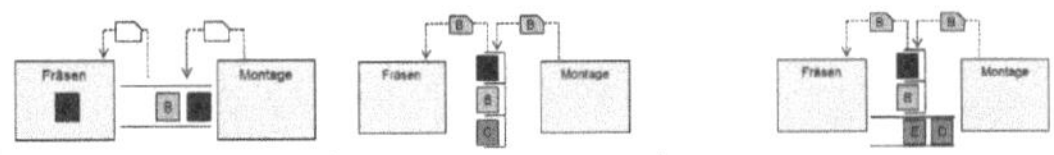

Criteria	FIFO	Supermarket	Mixed Pull
Production volume	low	high	
Product variance	high	low	
Demand	stable	variable	
Process stability	high	medium	
Product dimensions	small-large	small	
Shelf life of the products	short	long	
Availabilty	Medium/ by order	immediate	
Unit costs	high	low	

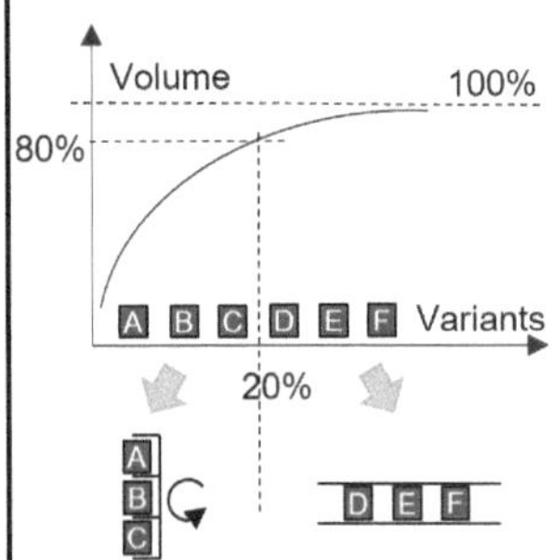

FIGURE 5.14 Criteria to choose suitable process connections.

5.2.3.4 Logistical Design of Process Connections

Once a suitable process connection has been selected, it must be implemented logistically to ensure the physical flow of items. The logistical design of process connections can be categorized into four transport options: workers' self-supply, milk run, direct transport, and rigid conveyor technology.

Worker self-supply involves employees within the processes independently procuring the required materials, either retrieving them from upstream processes or delivering them to the next step. However, this approach contradicts the fundamental separation of production and logistics activities and is only recommended for short-distance transports.

Properties of Source-sink Relationships	Areas of Application of Transport Forms and Value Stream Symbol			
	Workers' Self-Supply	Milkrun	Direct Traffic	Rigid Conveyor technology
Transport Route	Very short	Medium-length	Short Medium	Short Medium
Allowed Transport Time	Negligible	Long	Short	Short
Transport Unit (Size)	Small load carrier	Load carrier on transport rack	Load carriers on pallet	Suitable for automatic handling
Transport Unit (Weight)	Under 15 kg	Under 15 kg	Not Restricted	Not Restricted
Stability of Demand	Not Restricted	Regular Plannable	Irregular	Regular Plannable
Demand Frequency	High	High	Medium	Low

FIGURE 5.15 Criteria to choose suitable transport forms (Kaiser 2021).

Milk run systems operate by making regular deliveries to various stations in cyclical traffic along established routes.

Direct transport establishes a direct material transport link between two processes, maintaining a one-to-one relationship between material source and destination.

Rigid conveyor technology represents a distinctive case among these transport options. Here, processes are interconnected by technical systems such as conveyor belts, resulting in a highly inflexible setup.

Whenever possible, the transport system should be chosen to supply processes in accordance with standardized procedures, like fixed schedules and routes, to reduce waste and detect deviations. This is typically not the case with worker self-supply. For instance, the integration of customer-specific parts can be explored using self-navigating transport robots, often referred to as Automated Mobile Robots (AMR).

The following overview provides a summary of the appropriateness of transport systems based on the source-sink relationship of process connections, aiding in the selection of suitable transport systems (Figure 5.15).

5.2.4 Step 6: Determine Value Stream Control

In the final stage of classic value stream design, the control point for production orders needs to be determined. Ideally, orders should be controlled at a single point within the value stream, known as the pacemaker. This pacemaker separates the value stream into two segments: customer-neutral production processes (upstream) and order- or customer-specific production processes (downstream). Consequently, processes upstream of the pacemaker are regulated using a supermarket pull system and associated production kanban. The processes downstream of the pacemaker follow the FiFo principle, receiving the next product to be produced in the sequence defined at the pacemaker. The choice of the entry point for control is influenced by the capability of processes to follow the pacemaker's sequence without significant setup times. Simultaneously, the selected order processing principle must be considered.

For controlling a make-to-order value stream, the pacemaker should be placed far upstream. Subsequent processes are then connected through a FiFo chain, leading up to product delivery. A make-to-stock value stream, on the other hand, can typically be controlled from the finished goods warehouse via supermarket systems. Opting for a single control point in the value stream offers several advantages:

- It reduces coordination efforts, as all other processes are governed from the pacemaker.
- Work increments are authorized at the pacemaker for the entire value stream.
- Upstream processes only produce when the pacemaker sends a pull signal.
- Subsequent processes adhere to the sequence defined by the pacemaker.

Order release at the pacemaker should be managed in a way that avoids overload or under-load at processes. To achieve this, load monitoring or load forecasting should be carried out at the value stream's bottleneck. The next order should only be released at the pacemaker if there is free capacity at the bottleneck process. To ensure an even utilization of all processes, different product variants and volumes should be evenly distributed over a defined planning period and managed at the pacemaker.

To accommodate processes with unavoidable setup times in combining orders into batches, it's possible to work with the earliest start and latest delivery times for production orders. The provision of a time span between earliest start and latest delivery of a production order helps a workstation to reduce setup losses and compensate for cycle time variations while maintaining fixed throughput times.

As a result of steps three to six, a target concept for the product flow has been developed. All order processing processes are defined, designed, and interconnected through control elements.

OPPORTUNITIES FOR IMPROVEMENT THROUGH AUTOMATION IN THE INDIRECT AREA

"Automation" entails the substitution of human labor with machinery and systems. In manufacturing, this often involves the deployment of industrial robots, autonomous transport systems, or automated facilities. However, within indirect domains, particularly those encompassing a substantial share of routine tasks, it is crucial to explore the integration of automated solutions. This exploration can unlock potential improvements related to process stability, consistency in output quality, and reduced order processing times.

In the context of the indirect domain, automation signifies the delegation of information processing tasks from human workers to software. Adhering to the established principles of Value Stream Design 4.0 within the indirect sphere paves the way for structured automation opportunities. An essential prerequisite for this is the elimination of wasteful activities and the standardization of processes, encompassing clearly defined input and output information.

Processes that predominantly consist of routine work and exhibit limited creative elements are particularly amenable to automation. In the context of integration within Value Stream Design 4.0, it doesn't matter whether a process function or value creation is executed by software or by human personnel within a department. The fundamental input and output products of the process may remain unaltered; what matters is the corresponding transformation of material and information orchestrated by the process.

5.2.5 Step 7: Identify Information Needs for the Processes

Following the completion of the product flow design, this step involves the preparation for the interconnection of the value stream. In this phase, a thorough assessment is made to identify which processes within the value stream are linked to the information requirements established in Phase III, and how these requirements can be met.

To facilitate this evaluation, the information matrix is revisited and expanded upon. All previously outlined processes and process connections within the value stream are documented in the available columns under "Processes in the value stream". Subsequently, each of the defined information requirements from Phase I is scrutinized to determine which specific process must supply the necessary information to meet that requirement.

When a correlation between a defined information need and a process within the value stream is identified, it is denoted by a "cross" at the corresponding position, thereby illustrating the connection (Figure 5.16).

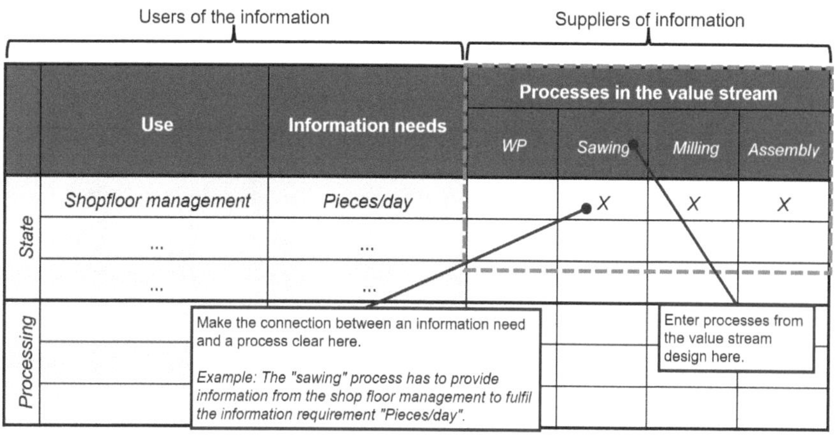

FIGURE 5.16 Information matrix to define connections between users and suppliers of information.

5.3 PHASE V: NETWORKING THE VALUE STREAM

Following the completion of Phase IV, the process image of the new value stream has been established. Logistical connections and their control have been defined. Hence, in the fifth and final phase, the value stream can now be integrated in terms of information logistics. This integration focuses on two types of information (as discussed in Section 4.1):

- Order processing information: This information is essential for processes to carry out their order fulfilment functions or is generated for other processes to perform their respective functions. The details for these are outlined in step 8.
- Status information: This information is required to execute the business model or by functional areas to meet their tasks and objectives. The requirements for status information were initially documented in Phase III and linked to the value stream processes. In step 9, they are further refined for each individual process.

Once all the necessary information inputs and outputs have been precisely defined, step 10 facilitates the ultimate connection of information throughout the entire value stream. The overarching goal of these three steps is to establish a continuous and waste-free information flow (Figure 5.17).

Method steps 8–10 should be applied to all processes within the value stream, commencing with the final process. Initiating the process with the last step allows for the systematic recording of information flows based on requirements, gradually establishing a network for information logistics throughout the entire value stream. The process culminates at the value stream's inception point: the customer. Even for the customer, it is essential to define the information they need to provide, which is indispensable for the value stream in the context of order processing. This is undertaken with the objective of minimizing queries and ensuring an efficient connection between the customer and the order processing system.

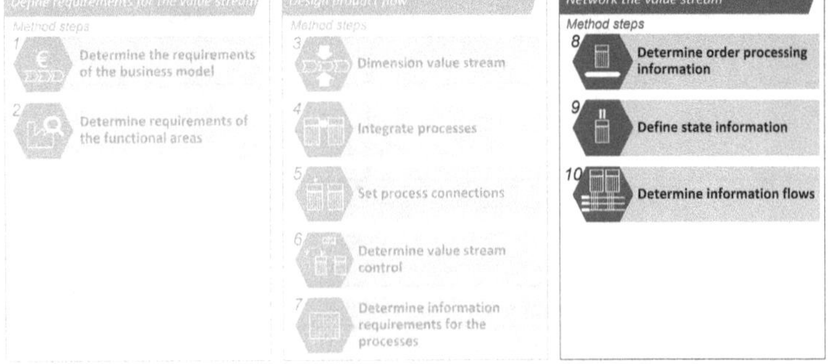

FIGURE 5.17 Value Stream Design 4.0 – Phase V.

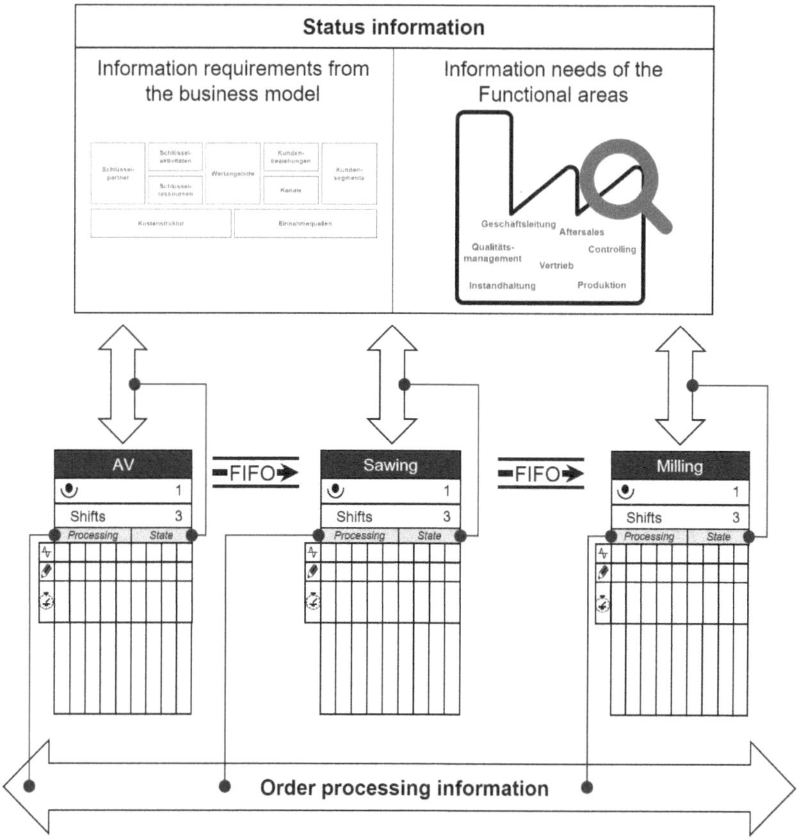

FIGURE 5.18 Concept of connecting horizontal and vertical information flows.

Figure 5.18 provides a concise overview of the design principles that serve as the basis for deriving information flows in Value Stream Design 4.0.

5.3.1 Step 8: Determine Order Processing Information

The objective of this methodological phase is to pinpoint and establish all essential information that processes necessitate for the execution of their functions, tasks, and activities within the value stream. To accomplish this, a comprehensive analysis of the processes in terms of their information prerequisites is imperative. These information requirements may encompass vital data such as order specifics or assembly instructions.

Furthermore, it's crucial to recognize that the information needs of these processes must be met by other processes (or by the customer). Consequently, this phase encompasses more than merely defining the information required by an individual process; it also entails specifying the information that the process must generate and provide to enable other processes to fulfil their respective tasks. The goal is simply to eliminate any waiting time due to missing or incomplete information. To encapsulate, we must address the following two fundamental questions:

What information does the process require to execute its functions and tasks within the scope of order processing?
What information does the process need to provide for other processes, enabling them to carry out their functions and tasks effectively?

The identified information requirements should be incorporated into the process box. In the provided example, the process necessitates order documents and the NC program code to carry out its functions effectively (Figure 5.19).

The information matrix serves as a tool for orchestrating the "horizontal" connectivity within the value stream. It entails assigning the information needs of a process to their corresponding sources. To achieve this, the process, along with its information requirements, is integrated into the rows of the information matrix and connected to the providing process using an "X".

In the presented example, the information requirements of the milling process are generated by the Work Preparation Department (WP) department. These requirements are illustrated in the information matrix, as depicted in the figure (Figure 5.20).

Next, you need to identify the information that the process must supply to other processes within the value stream. This information can be extracted from the information matrix. The "X" marks in the column corresponding to the specific process indicate the required information for supply. These details should be integrated into the process box within the value stream map.

However, in the case of the previously examined milling process, it doesn't serve as an information supplier for other processes. There are no "X" links present in the information matrix column associated with this process (Figure 5.21).

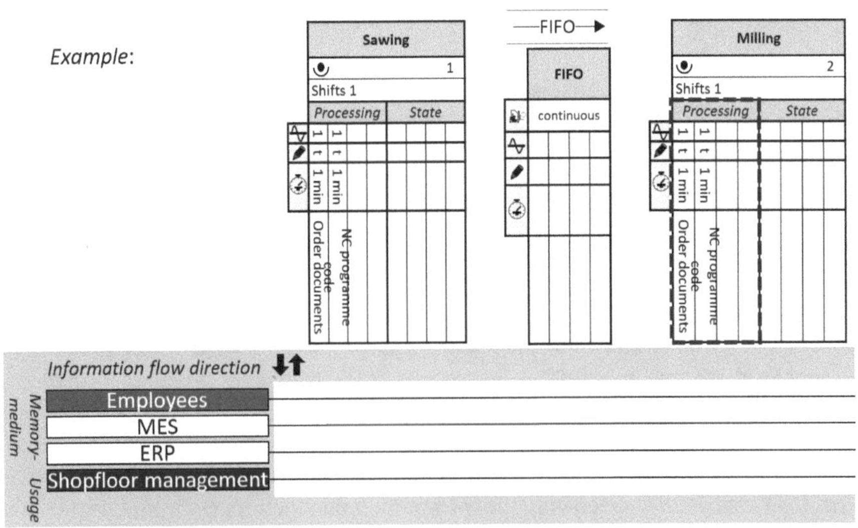

FIGURE 5.19 Population of the process box with required order processing information – e.g., milling.

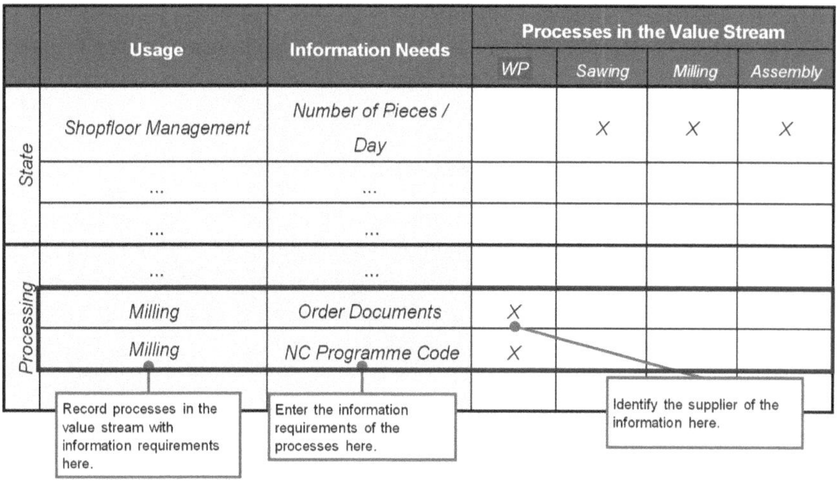

FIGURE 5.20 Linking requirements of each process with information suppliers in the information matrix.

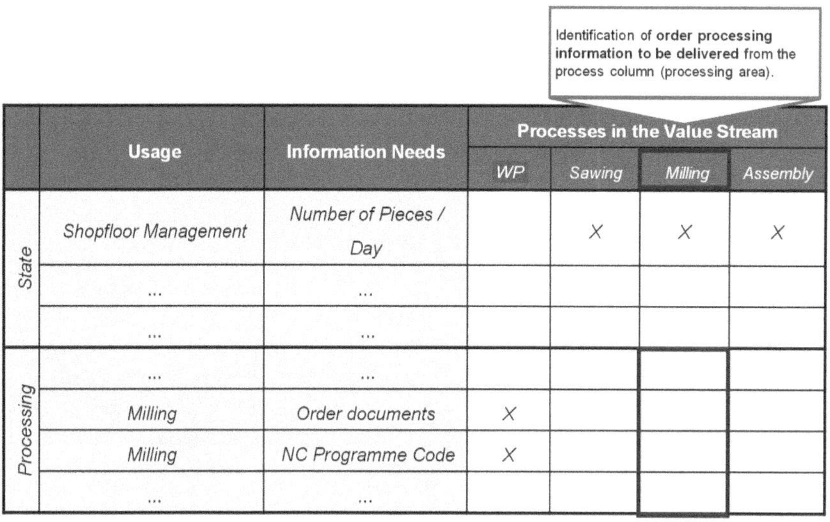

FIGURE 5.21 Linking of information supply by a process to other processes in the information matrix.

To illustrate the process of identifying the information that needs to be supplied during order processing, let's consider the WP process. If this process has been properly designed in the subsequent method application, you will find "X" marks in the corresponding column. These "X" marks signify that the creation of order documents and the generation of the NC program code for the milling process are required. The information that has been pinpointed in this manner should then be integrated into the process box of the WP process within the value stream map.

After completing this method step for all processes within the value stream, it's crucial to perform a final check to ensure that all identified information requirements align with the corresponding information provisions.

5.3.2 Step 9: Define Status Information

Following this, the design of the status information flow is undertaken. Requirements for status information originate from the business model or within functional areas, such as shop floor management, and have already been integrated during Phase III.

Within the scope of this methodological step, the process currently under review is evaluated in response to the following guiding question:

> What information does the process need to generate and provide in order to meet the information logistics requirements outlined by the business model and functional areas?

The state information that needs to be provided can be extracted from the respective process column in the information matrix. These details are then incorporated into the process box within the value stream map, specifically in the status information area.

For instance, in the example, the milling process needs to provide the number of parts produced for shop floor management (Figures 5.22 and 5.23).

Identification of **state information to be supplied** from the process column (area state).

	Usage	Information Needs	Processes in the Value Stream			
			WP	Sawing	Milling	Assembly
State	Shopfloor Management	Number of Pieces/Day		X	X	X
	...	...				
	...	...				
Processing	...	...				
	Milling	Order Documents	X			
	Milling	NC Programme Code	X			
	...	...				

FIGURE 5.22 Information supply by a process to functional areas and in support of the business model.

Example:

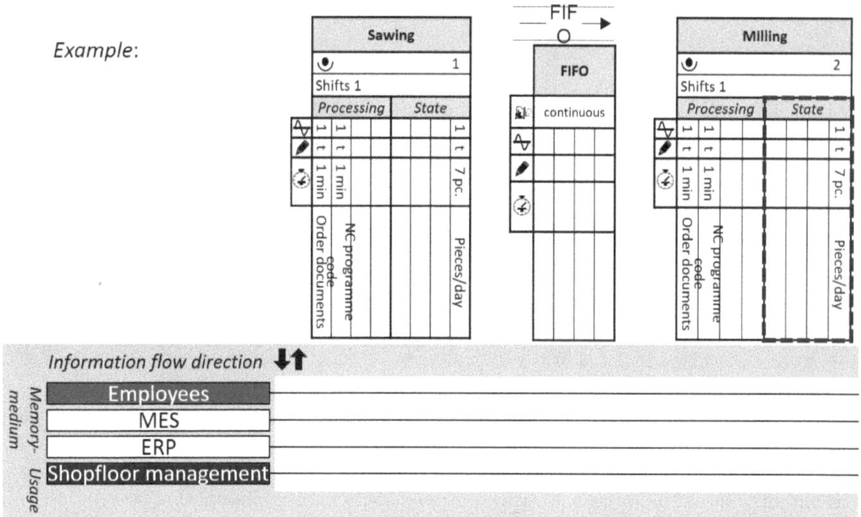

FIGURE 5.23 Population of the process box for required information about the process state – e.g., milling.

5.3.3 Step 10: Determine Information Flows

Now, the defined information flows between individual processes can be interlinked within the value stream map. This linkage is accomplished through storage media and IT systems, with the aim of efficiently and minimizing waste in connecting information supply with information demand.

In the value stream map, the specified information requirements and provisions of the process are linked to the corresponding storage medium's horizontal line, which is used for retrieval or storage. A node is placed at the intersection of these lines. Each vertical connecting line thus represents either information retrieval or information provision. If a suitable storage medium or IT system isn't defined or available within the company, introducing a new system may be necessary.

Additionally, it's essential to determine the purpose of each piece of information. Every item of information should have a clear and practical use. The utilization of an item of information is indicated by a vertical dotted line connecting the storage medium to its purpose. If the necessary usage line is absent in the value stream map, it should be added.

With the completion of this final step 10, the processes in the value stream are integrated in terms of information logistics. When this step is executed for all processes, a new target image for the value stream emerges. The now available value stream map forms the foundation for the subsequent realization of the value stream (Figure 5.24).

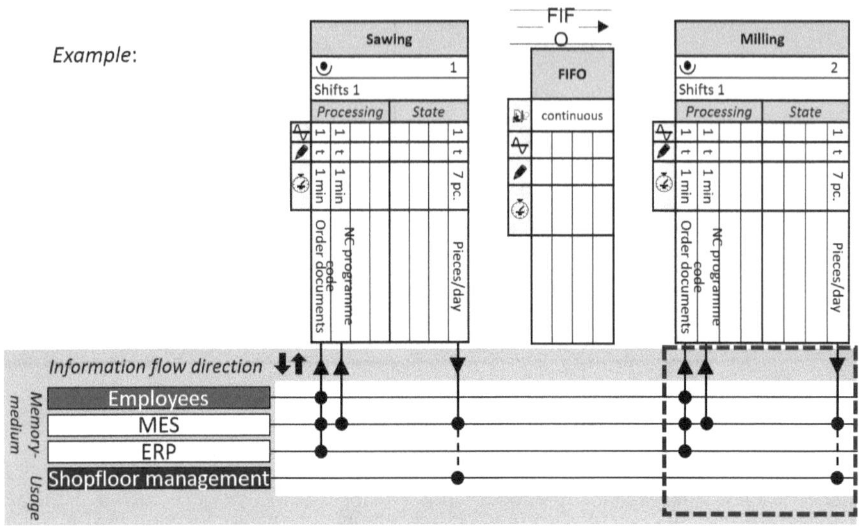

FIGURE 5.24 Connection of information sources and sinks in the VSD diagram.

5.3.4 Practical Example – Application of Value Stream Design 4.0- Pumpen AG

In the course of developing its business model, Pumpen AG has made a strategic decision to exclusively accept orders through a specially designed online portal in the future. This portal is seamlessly integrated with configuration software, empowering customers to independently create their products. The advantage for customers lies in the guided product planning process, where they are prompted to provide all necessary information. Consequently, feedback on the feasibility of their desired product can be offered immediately. For Pumpen AG, this approach eliminates coordination efforts and ensures that all essential information is collected and transmitted in a standardized manner, enabling efficient and automated processing within the company.

When delineating the functional area requirements, it becomes evident that at Pumpen AG, the primary information recipients are shop floor management and quality management. Both sectors require data from the direct areas concerning setup times, quantities, rejects, and cycle times. Additionally, an OEE (Overall Equipment Effectiveness) should be recorded for the turning process. In the indirect areas, the recording should encompass the number of units and cycle times.

Within Pumpen AG's product flow, processes can be integrated into both the direct and indirect areas, allowing the implementation of FIFO systems as process connections. In the direct area, this results in a streamlined process responsible for both assembly and product packaging. Meanwhile, in the indirect area, process integration is chiefly achieved through the use of configuration software, significantly

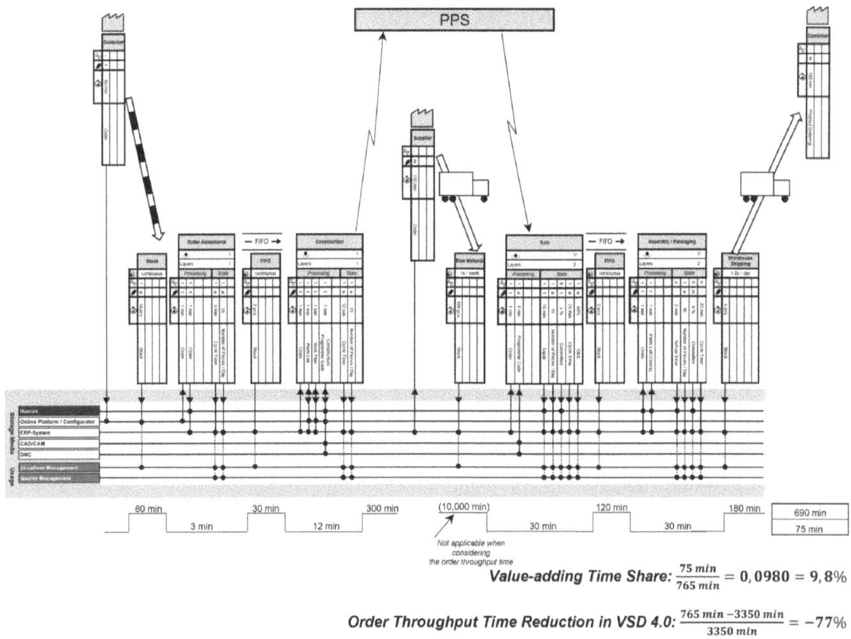

FIGURE 5.25 VSD diagram for Pumpen AG.

reducing manual programming efforts. These can be integrated into a newly designed construction process through automated processing of the construction data sourced from orders.

Regarding the design of information flows, Pumpen AG has recognized significant potential for improvement due to the existence of diverse systems for ordering products and a lack of interfaces between various storage media, resulting in manual information transfers. In the future, orders will be seamlessly transferred and processed via well-defined interfaces between storage media. The design department employs the customer-configured components to generate individual program codes for the turning process and corresponding parts lists for assembly through the use of a configurator. A CAD/CAM system has been introduced for this purpose. The production processes can directly access order data and parts lists from the ERP (Enterprise Resource Planning) system, eliminating the need for paper documents. Furthermore, the ERP system supplies the required information to meet the needs of functional areas. The resulting improvements are detailed on the following page (Figure 5.25).

5.4 WHAT HAS BEEN ACHIEVED WITH VALUE STREAM DESIGN 4.0?

Development of a New Target State

This target state encompasses both the entire material flow and the belonging information flows. It delineates all the constituents of a value stream.

Connecting the Business Model and the Value Stream

The business model was considered throughout the value stream design. This approach allows the value stream to potentially yield a strategic competitive advantage or facilitate the implementation of process requirements via innovative business models.

Reducing Waste in Information Logistics

The systematic approach and the benefit-oriented definition of information flows play a pivotal role in minimizing and preventing waste in information logistics processes.

Commencement of Implementation

The defined target state serves as the starting point for value stream planning, providing a foundation upon which subsequent projects can be delineated and further details can be elaborated.

Utilization as a Communication Tool

The presentation of the new target state can serve as an effective communication tool. It provides a foundation for discussing and comparing various design alternatives.

Harnessing the Value of Information

Information is strategically employed to generate value for both the customer and internal operations.

Unlocking the Potential of Digitalization and Process Networking

The potential benefits arising from the digitization and networking of processes can be effectively visualized through graphical representation. This provides a clear understanding of the specific information requirements and their corresponding areas of application.

NOTES

1 Bieger and Reinhold (2011).
2 Wiegand and Pöhls (2009); Magenheimer (2014); Hartmann (2022).
3 Erlach (2010).

REFERENCES

Bieger, Thomas; Reinhold, Stephan (2011): Das wertbasierte Geschäftsmodell - Ein aktuali-sierter Strukturierungsansatz. In: Bieger, Knyphausen-Aufseß, Krys (eds.): *Innovative Ge-schäftsmodelle*; Berlin, Heidelberg: Springer.

Erlach, Klaus (2010): *Wertstromdesign - Der Weg zur schlanken Fabrik*; 2., edit. und erw. ed.; Berlin, Heidelberg: Springer.

Hartmann, Lukas (2022): *Wertstromdesign 4.0- Methode zur integrierten Gestaltung von Material- und Informationsflüssen für schlanke Wertströme*; Düren: Shaker Verlag.

Kaiser, Joscha (2021): *Logistische Planungsalternativen im Wertstromdesign - Ein Ansatz zur Identifikation und Auswahl von Soll-Zuständen in der Materialflussgestaltung innerbetrieblicher Wertströme*; Düren: Shaker.

Magenheimer, Kai Alexander (2014): *Lean Management in Indirect Business Units - Modelling, Analysis and Evaluation of Waste*; Munich.

Osterwalder, Alexander; Pigneur, Yves (2011): *Business Model Generation*; Frankfurt: Cam-pus.

Wiegand, Bodo; Pöhls, Katja (2009): *Lean Administration II - How to Manage Business Pro-cesses Properly; Step 2: The Optimisation; Version 2.0*; Aachen: Lean Management Institut

Appendix and Symbols

A.1 KEY FIGURES FOR INFORMATION LOGISTICS WASTE ANALYSIS

Using VSA4.0 and VSD4.0, one can analyze, design, and compare information flows effectively. This allows for the quantification of enhancements in a value stream's information flows. To achieve this, we suggest three key metrics, detailed as follows: the digitization rate, data accessibility, and data utilization.[1]

A.1.1 Advanced Value Stream Metrics – Digitization Rate

Using this metric, all information flows within the value stream are assessed in terms of direct digital capture and processing. The goal is to determine if information can be processed without any involvement of analogue storage media. To achieve this, every instance of information processing, including storage and retrieval, within the value stream is scrutinized, along with their respective key performance indicators (KPIs). When information flows within an element are entirely digitally processed, they are classified as a "digital information point" (refer to the numerator in the formula). Information flows that currently involve interactions either with digital or with analogue storage media are grouped under "all information points" (found in the denominator of the formula). The digitization rate is calculated as the ratio of "digital information points" to "all information points". This KPI will always be less than or equal to 100%. Two perspectives can be applied: the digitization rate can be calculated for the entire value stream or for each individual process or logistics element.

A.1.1.1 For the Entire Value Stream

Examining the digitization rate from a value stream perspective provides an overview of the entire value stream. It allows for comparisons between different value streams, aiding in the identification of best practices for future improvement projects. It can also be used to assess improvements following the implementation of Value Stream Design 4.0 (VSD4.0).

A.1.1.2 For Individual Processes and Logistics Elements

The digitization rate is calculated separately for each process and logistics element. Sorting the values in ascending order helps in identifying the potential for digitization (bearing in mind the cost benefit ratio) (Figure A.1).

Example:

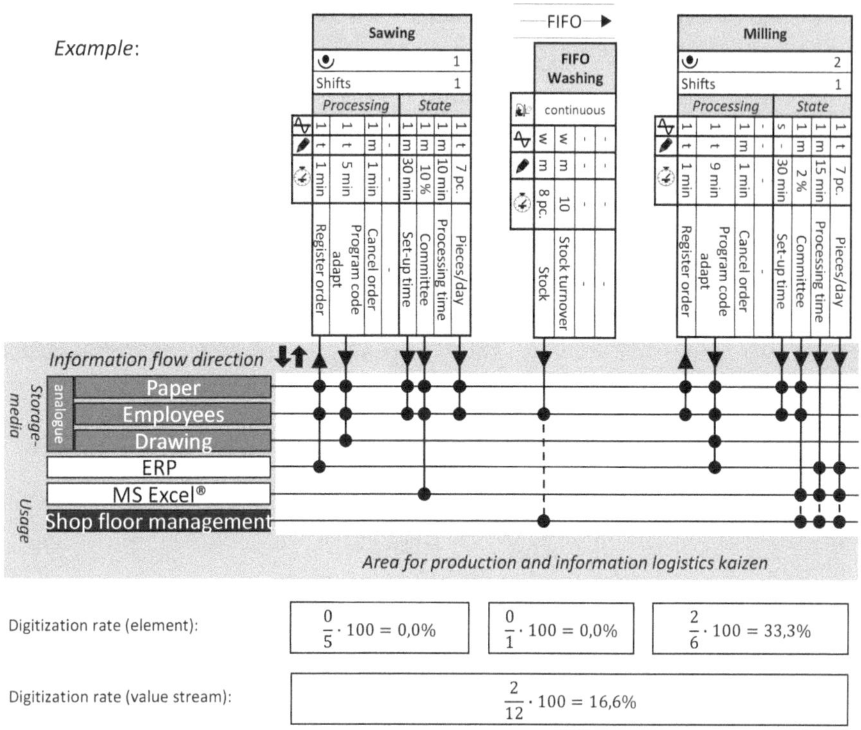

FIGURE A.1 Calculation example digitization rate.

On the following page, you will find the formulas and an illustrative example.

$$\text{Digitisationrate}(\text{Element}_x) = \frac{\sum_1^n \text{Digital Information Point}_i}{\sum_1^n \text{All Information Points}_i} * 100\ [\%]$$

$$\text{Digitisationrate\_Value\_Stream} = \frac{\sum_1^m \text{Digitisationrate}(\text{Element}_x)}{\sum_1^m (\text{Element}_x)} [\%]$$

with:

- Element: Process step or logistics element;
- Digital information point: Information, including key performance indicators (KPIs) and recorded activities within a value stream element, is deemed to be digitally processed when it doesn't incur a disruption due to analogue storage media.
- All information points: Information in the form of KPIs and activities is recorded within a value stream element, independent of the storage medium used.

A.1.2 Advanced Value Stream Metrics – Information Availability

The metric of information availability assesses the actual recording of KPIs as defined by the value stream team within your value stream's processes and logistics elements. However, only KPIs that are part of the defined set (KPI planned) are considered for this calculation. This approach prevents the inclusion of arbitrary KPIs solely to inflate the information availability percentage.

It's important to note that this metric remains unaffected by the number of storage media employed. Consequently, it serves as a valuable tool for evaluating whether and to what extent the designated KPIs are being captured within a given value stream element. A high percentage of information availability can facilitate continuous improvement efforts within companies and promote problem-oriented decision-making processes (Figure A.2).

$$\text{Information Availability} - \text{Process\_}x = \frac{\sum_{1}^{n} \text{KPIs\_recorded}_i}{\sum_{1}^{n} \text{KPIs\_planned}_i} * 100[\%]$$

with:

- KPIs_recorded: KPIs actually recorded within the process;
- KPIs_planned: Number of KPIs to be recorded within the process (previously defined in step 1 of the VSA4.0 method)

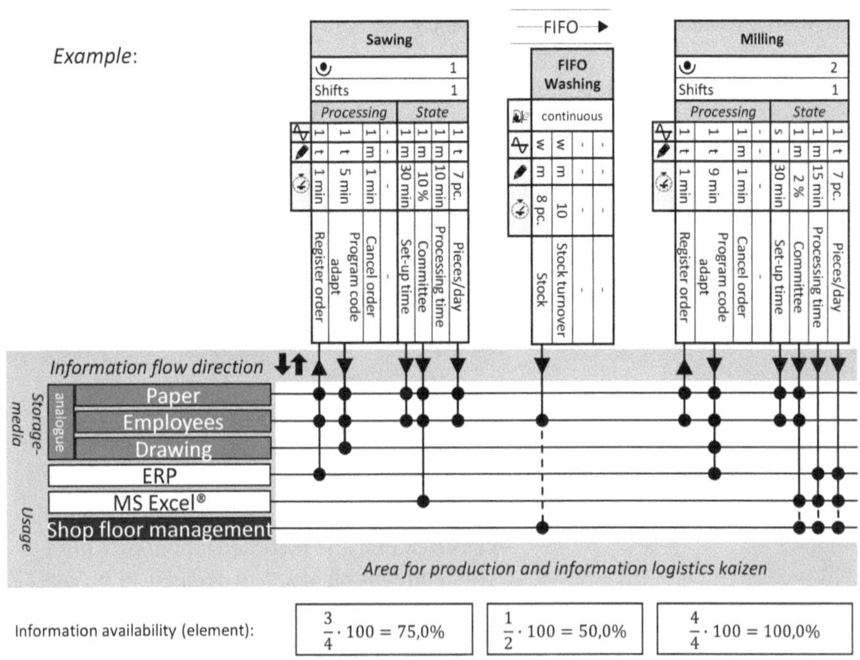

FIGURE A.2 Calculation example information availability.

A.1.3 Advanced Value Stream Metrics – Information Utilization

This metric, linked to information availability, gauges how effectively the recorded KPIs are actively applied. Examples of information use include shop floor management, automated decision-making, and documentation. Multiple uses of the same information point are counted once in the metric calculation to prevent misinterpretation. Low values suggest potential shortcomings in activities like shop floor management or its associated improvement initiatives (Figures A.3).

$$\text{Information Usage}(\text{Element}\_x) = \frac{\sum_{1}^{n} \text{KPI}\_\text{recorded \& used}_i}{\sum_{1}^{n} \text{KPI}\_\text{planned}_i} * 100[\%]$$

with

- Element: Process step or logistics element;
- KPI recorded and used: KPIs from the set of planned KPIs of the value stream element that is recorded and used;
- Key figure planned: Key figures to be recorded for a value stream element (defined in step 1).

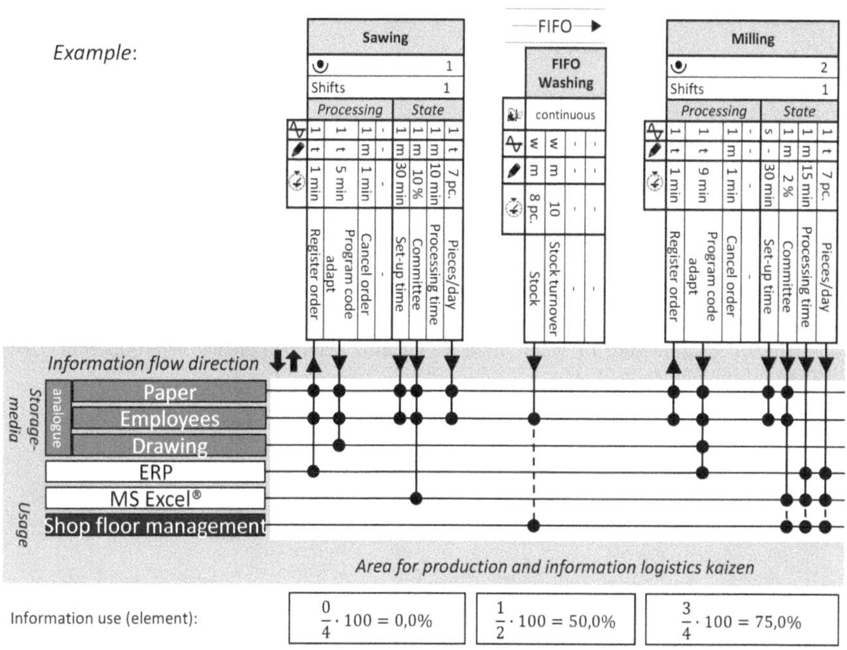

FIGURE A.3 Calculation example information utilizations.

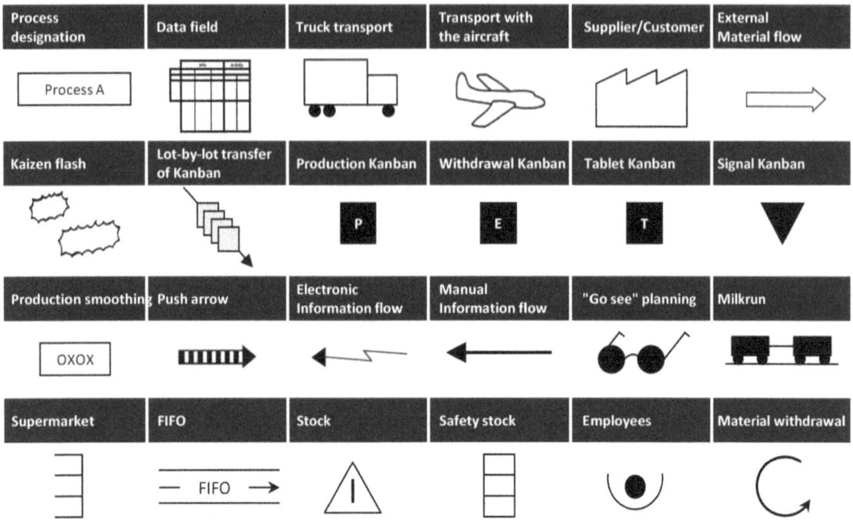

FIGURE A.4 Overview of value stream symbols.

A.2 VALUE STREAM SYMBOLS

See Figure A.4

NOTE

1 Here and in the following: Meudt (2020); Meudt, Metternich, Abele (2017).

REFERENCES

Meudt, Tobias (2020): *Wertstromanalyse 4.0- Eine Methode zur integrierten Erfassung und Analyse von Material- und Informationsflüssen in Wertströmen*; Düren: Shaker Verlag.

Meudt, Tobias; Metternich, Joachim; Abele, Eberhardt (2017): Value stream mapping 4.0-Holistic examination of value stream and information logistics in production; *CIRP Annals* 66 (1), pp. 413–416.

Index

Kolibri 360 GmbH
Ferdinand-Jühlke-Straße 7
99095 Erfurt, Deutschland
produktsicherheit@kolibri360.de